Introduction to MATLAB® 6

Second Edition

Delores M. Etter and
David C. Kuncicky with Doug Hull

PEARSON

Prentice
Hall

Pearson Education, Inc.
Upper Saddle River, NJ 07458

Library of Congress Cataloging-in-Publication Data

Etter, D. M.
    Introduction to MATLAB 6 / Delores M. Etter and David C. Kuncicky with Doug Hull.—
2nd ed.
        p. cm. —(The Prentice Hall engineering source)
    Includes bibliographical references and index.
    ISBN 0-13-140918-2
      1. Engineering mathematics—Data processing. 2. MATLAB. I. Kuncicky, David C. II.
Hull, Douglas W. III. Title. IV. ESource—the Prentice Hall engineering source.

TA345.E8725 2003
620'.001'51—dc21

2003046362

Vice President and Editorial Director, ECS: Marcia J. Horton
Executive Editor: Eric Svendsen
Associate Editor: Dee Bernhard
Vice President and Director of Production and Manufacturing, ESM: David W. Riccardi
Executive Managing Editor: Vince O'Brien
Managing Editor: David A. George
Production Editor: Barbara A. Till
Director of Creative Services: Paul Belfanti
Creative Director: Carole Anson
Art Director: Jayne Conte
Art Editor: Greg Dulles
Manufacturing Manager: Trudy Pisciotti
Manufacturing Buyer: Lisa McDowell
Marketing Manager: Holly Stark

 © 2004, 2002 by Pearson Education, Inc.
Pearson Prentice Hall
Pearson Education, Inc.
Upper Saddle River, NJ 074588

Printed in the United States of America.

10 9 8 7 6 5 4 3 2 1

ISBN 0-13-140918-2

Pearson Education Ltd., *London*
Pearson Education Australia Pty. Ltd., *Sydney*
Pearson Education Singapore, Pte. Ltd.
Pearson Education North Asia Ltd., *Hong Kong*
Pearson Education Canada, Inc., *Toronto*
Pearson Educación de Mexico, S.A. de C.V.
Pearson Education—Japan, *Tokyo*
Pearson Education Malaysia, Pte. Ltd.
Pearson Education, *Upper Saddle River, New Jersey*

About ESource

ESource—The Prentice Hall Engineering Source—
www.prenhall.com/esource

ESource—The Prentice Hall Engineering Source gives professors the power to harness the full potential of their text and their first-year engineering course. More than just a collection of books, ESource is a unique publishing system revolving around the ESource website—www.prenhall.com/esource. ESource enables you to put your stamp on your book just as you do your course. It lets you:

Control   You choose exactly what chapters are in your book and in what order they appear. Of course, you can choose the entire book if you'd like and stay with the authors' original order.

Optimize   Get the most from your book and your course. ESource lets you produce the optimal text for your students needs.

Customize   You can add your own material anywhere in your text's presentation, and your final product will arrive at your bookstore as a professionally formatted text. Of course, all titles in this series are available as stand-alone texts, or as bundles of two or more books sold at a discount. Contact your PH sales rep for discount information.

ESource ACCESS

Professors who choose to bundle two or more texts from the ESource series for their class, or use an ESource custom book will be providing their students with an on-line library of intro engineering content—ESource Access. We've designed ESource ACCESS to provide students a flexible, searchable, on-line resource. Free access codes come in bundles and custom books are valid for one year after initial log-on. Contact your PH sales rep for more information.

ESource Content

All the content in ESource was written by educators specifically for freshman/first-year students. Authors tried to strike a balanced level of presentation, an approach that was neither formulaic nor trivial, and one that did not focus too heavily on advanced topics that most introductory students do not encounter until later classes. Because many professors do not have extensive time to cover these topics in the classroom, authors prepared each text with the idea that many students would use it for self-instruction and independent study. Students should be able to use this content to learn the software tool or subject on their own.

While authors had the freedom to write texts in a style appropriate to their particular subject, all followed certain guidelines created to promote a consistency that makes students comfortable. Namely, every chapter opens with a clear set of **Objectives**, includes **Practice Boxes** throughout the chapter, and ends with a number of **Problems**, and a list of **Key Terms**. **Applications Boxes** are spread throughout the book with the intent of giving students a real-world perspective of engineering. **Success Boxes** provide the student with advice about college study skills, and help students avoid the common pitfalls of first-year students. In addition, this series contains an entire book titled *Engineering Success* by Peter Schiavone of the University of Alberta intended to expose students quickly to what it takes to be an engineering student.

Creating Your Book

Using ESource is simple. You preview the content either on-line or through examination copies of the books you can request on-line, from your PH sales rep, or by calling 1-800-526-0485. Create an on-line outline of the content you want, in the order you want, using ESource's simple interface. Insert your own material into the text flow. If you are not ready to order, ESource will save your work. You can come back at any time and change, re-arrange, or add more material to your creation. Once you're finished you'll automatically receive an ISBN. Give it to your bookstore and your book will arrive on their shelves four to six weeks after they order. Your custom desk copies with their instructor supplements will arrive at your address at the same time.

To learn more about this new system for creating the perfect textbook, go to www.prenhall.com/esource. You can either go through the on-line walkthrough of how to create a book, or experiment yourself.

Supplements

Adopters of ESource receive an instructor's CD that contains professor and student code from the books in the series, as well as other instruction aides provided by authors. The website also holds approximately **350 PowerPoint transparencies** created by Jack Leifer of University of Kentucky–Paducah. Professors can either follow these transparencies as pre-prepared lectures or use them as the basis for their own custom presentations.

Titles in the ESource Series

Design Concepts for Engineers, 2/e
0-13-093430-5
Mark Horenstein

Engineering Success, 2/e
0-13-041827-7
Peter Schiavone

Engineering Design and Problem Solving, 2E
ISBN 0-13-093399-6
Steven K. Howell

Exploring Engineering
ISBN 0-13-093442-9
Joe King

Engineering Ethics
0-13-784224-4
Charles B. Fleddermann

Introduction to Engineering Analysis
0-13-016733-9
Kirk D. Hagen

Introduction to Engineering Experimentation
0-13-032835-9
Ronald W. Larsen, John T. Sears, and Royce Wilkinson

Introduction to Mechanical Engineering
0-13-019640-1
Robert Rizza

Introduction to Electrical and Computer Engineering
0-13-033363-8
Charles B. Fleddermann and Martin Bradshaw

Introduction to MATLAB 6—Update
0-13-140918-2
Delores Etter and David C. Kuncicky, with Douglas W. Hull

MATLAB Programming
0-13-035127-X
David C. Kuncicky

Introduction to MATLAB
0-13-013149-0
Delores Etter with David C. Kuncicky

Introduction to Mathcad 2000
0-13-020007-7
Ronald W. Larsen

Introduction to Mathcad 11
0-13-008177-9
David W. Larsen

Introduction to Maple 8
0-13-032844-8
David I. Schwartz

Mathematics Review
0-13-011501-0
Peter Schiavone

Power Programming with VBA/Excel
0-13-047377-4
Steven C. Chapra

Introduction to Excel 2002
0-13-008175-2
David C. Kuncicky

Engineering with Excel
ISBN 0-13-017696-6
Ronald W. Larsen

Introduction to Word 2002
0-13-008170-1
David C. Kuncicky

Introduction to PowerPoint 2002
0-13-008179-5
Jack Leifer

Graphics Concepts
0-13-030687-8
Richard M. Lueptow

Graphics Concepts with SolidWorks 2/e
0-13-140915-8
Richard M. Lueptow and Michael Minbiole

Graphics Concepts with Pro/ENGINEER
0-13-014154-2
Richard M. Lueptow, Jim Steger, and
Michael T. Snyder

Introduction to AutoCAD 2000
0-13-016732-0
Mark Dix and Paul Riley

Introduction to AutoCAD, R. 14
0-13-011001-9
Mark Dix and Paul Riley

Introduction to UNIX
0-13-095135-8
David I. Schwartz

Introduction to the Internet, 3/e
0-13-031355-6
Scott D. James

Introduction to Visual Basic 6.0
0-13-026813-5
David I. Schneider

Introduction to C
0-13-011854-0
Delores Etter

Introduction to C++
0-13-011855-9
Delores Etter

Introduction to FORTRAN 90
0-13-013146-6
Larry Nyhoff and Sanford Leestma

Introduction to Java
0-13-919416-9
Stephen J. Chapman

About the Authors

No project could ever come to pass without a group of authors who have the vision and the courage to turn a stack of blank paper into a book. The authors in this series, who worked diligently to produce their books, provide the building blocks of the series.

Martin D. Bradshaw was born in Pittsburg, KS in 1936, grew up in Kansas and the surrounding states of Arkansas and Missouri, graduating from Newton High School, Newton, KS in 1954. He received the B.S.E.E. and M.S.E.E. degrees from the University of Wichita in 1958 and 1961, respectively. A Ford Foundation fellowship at Carnegie Institute of Technology followed from 1961 to 1963 and he received the Ph.D. degree in electrical engineering in 1964. He spent his entire academic career with the Department of Electrical and Computer Engineering at the University of New Mexico (1961-1963 and 1991-1996). He served as the Assistant Dean for Special Programs with the UNM College of Engineering from 1974 to 1976 and as the Associate Chairman for the EECE Department from 1993 to 1996. During the period 1987-1991 he was a consultant with his own company, EE Problem Solvers. During 1978 he spent a sabbatical year with the State Electricity Commission of Victoria, Melbourne, Australia. From 1979 to 1981 he served an IPA assignment as a Project Officer at the U.S. Air Force Weapons Laboratory, Kirkland AFB, Albuquerque, NM. He has won numerous local, regional, and national teaching awards, including the George Westinghouse Award from the ASEE in 1973. He was awarded the IEEE Centennial Medal in 2000.

Acknowledgments: Dr. Bradshaw would like to acknowledge his late mother, who gave him a great love of reading and learning, and his father, who taught him to persist until the job is finished. The encouragement of his wife, Jo, and his six children is a never-ending inspiration.

Stephen J. Chapman received a B.S. degree in Electrical Engineering from Louisiana State University (1975), the M.S.E. degree in Electrical Engineering from the University of Central Florida (1979), and pursued further graduate studies at Rice University. Mr. Chapman is currently Manager of Technical Systems for British Aerospace Australia, in Melbourne, Australia. In this position, he provides technical direction and design authority for the work of younger engineers within the company. He also continues to teach at local universities on a part-time basis.

Mr. Chapman is a Senior Member of the Institute of Electrical and Electronics Engineers (and several of its component societies). He is also a member of the Association for Computing Machinery and the Institution of Engineers (Australia).

Steven C. Chapra presently holds the Louis Berger Chair for Computing and Engineering in the Civil and Environmental Engineering Department at Tufts University. Dr. Chapra received engineering degrees from Manhattan College and the University of Michigan. Before joining the faculty at Tufts, he taught at Texas A&M University, the University of Colorado, and Imperial College, London. His research interests focus on surface water-quality modeling and advanced computer applications in environmental engineering. He has published over 50 refereed journal articles, 20 software packages and 6 books. He has received a number of awards including the 1987 ASEE Merriam/Wiley Distinguished Author Award, the 1993 Rudolph Hering Medal, and teaching awards from Texas A&M, the University of Colorado, and the Association of Environmental Engineering and Science Professors.

Acknowledgments: To the Berger Family for their many contributions to engineering education. I would also like to thank David Clough for his friendship and insights, John Walkenbach for his wonderful books, and my colleague Lee Minardi and my students Kenny William, Robert Viesca and Jennifer Edelmann for their suggestions.

Mark Dix began working with AutoCAD in 1985 as a programmer for CAD Support Associates, Inc. He helped design a system for creating estimates and bills of material directly from AutoCAD drawing databases for use in the automated conveyor industry. This system became the basis for systems still widely in use today. In 1986 he began collaborating with Paul Riley to create AutoCAD training materials, combining Riley's background in industrial design and training with Dix's background in writing, curriculum development, and programming. Mr. Dix received the M.S. degree in education from the University of Massachusetts. He is currently the Director of Dearborn Academy High School in Arlington, Massachusetts.

Delores M. Etter is a Professor of Electrical and Computer Engineering at the University of Colorado. Dr. Etter was a faculty member at the University of New Mexico and also a Visiting Professor at Stanford University. Dr. Etter was responsible for the Freshman Engineering Program at the University of New Mexico and is active in the Integrated Teaching Laboratory at the University of Colorado. She was elected a Fellow of the Institute of Electrical and Electronics Engineers for her contributions to education and for her technical leadership in digital signal processing.

Charles B. Fleddermann is a professor in the Department of Electrical and Computer Engineering at the University of New Mexico in Albuquerque, New Mexico. All of his degrees are in electrical engineering: his Bachelor's degree from the University of Notre Dame, and the Master's and Ph.D. from the University of Illinois at Urbana-Champaign. Prof. Fleddermann developed an engineering ethics course for his department in response to the ABET requirement to incorporate ethics topics into the undergraduate engineering curriculum. *Engineering Ethics* was written as a vehicle for presenting ethical

theory, analysis, and problem solving to engineering undergraduates in a concise and readily accessible way.

Acknowledgments: I would like to thank Profs. Charles Harris and Michael Rabins of Texas A & M University whose NSF sponsored workshops on engineering ethics got me started thinking in this field. Special thanks to my wife Liz, who proofread the manuscript for this book, provided many useful suggestions, and who helped me learn how to teach "soft" topics to engineers.

Kirk D. Hagen is a professor at Weber State University in Ogden, Utah. He has taught introductory-level engineering courses and upper-division thermal science courses at WSU since 1993. He received his B.S. degree in physics from Weber State College and his M.S. degree in mechanical engineering from Utah State University, after which he worked as a thermal designer/analyst in the aerospace and electronics industries. After several years of engineering practice, he resumed his formal education, earning his Ph.D. in mechanical engineering at the University of Utah. Hagen is the author of an undergraduate heat transfer text.

Mark N. Horenstein is a Professor in the Department of Electrical and Computer Engineering at Boston University. He has degrees in Electrical Engineering from M.I.T. and U.C. Berkeley and has been involved in teaching engineering design for the greater part of his academic career. He devised and developed the senior design project class taken by all electrical and computer engineering students at Boston University. In this class, the students work for a virtual engineering company developing products and systems for real-world engineering and social-service clients.

Acknowledgments: I would like to thank Prof. James Bethune, the architect of the Peak Performance event at Boston University, for his permission to highlight the competition in my text. Several of the ideas relating to brainstorming and teamwork were derived from a workshop on engineering design offered by Prof. Charles Lovas of Southern Methodist University. The principles of estimation were derived in part from a freshman engineering problem posed by Prof. Thomas Kincaid of Boston University.

Steven Howell is the Chairman and a Professor of Mechanical Engineering at Lawrence Technological University. Prior to joining LTU in 2001, Dr. Howell led a knowledge-based engineering project for Visteon Automotive Systems and taught computer-aided design  classes for Ford Motor Company engineers. Dr. Howell also has a total of 15 years experience as an engineering faculty member at Northern Arizona University, the University of the Pacific, and the University of Zimbabwe. While at Northern Arizona University, he helped develop and implement an award-winning interdisciplinary series of design courses simulating a corporate engineering-design environment.

Douglas W. Hull is a graduate student in the Department of Mechanical Engineering at Carnegie Mellon University in Pittsburgh, Pennsylvania. He is the author of *Mastering Mechanics I Using Matlab 5*, and contributed to *Mechanics of Materials* by Bedford and Liechti. His  research in the Sensor Based Planning lab involves motion planning for hyper-redundant manipulators, also known as serpentine robots.

Scott D. James is a staff lecturer at Kettering University (formerly GMI Engineering & Management Institute) in Flint, Michigan. He is currently pursuing a Ph.D. in Systems Engineering with an emphasis on software engineering and computer-integrated manufac-  turing. He chose teaching as a profession after several years in the computer industry. "I thought that it was really important to know what it was like outside of academia. I wanted to provide students with classes that were up to date and provide the information that is really used and needed."

Acknowledgments: Scott would like to acknowledge his family for the time to work on the text and his students and peers at Kettering who offered helpful critiques of the materials that eventually became the book.

Joe King received the B.S. and M.S. degrees from the University of California at Davis. He is a Professor of Computer Engineering at the University of the Pacific, Stockton, CA, where he teaches courses in digital design, computer design, artificial intelligence, and com-  puter networking. Since joining the UOP faculty, Professor King has spent yearlong sabbaticals teaching in Zimbabwe, Singapore, and Finland. A licensed engineer in the state of California, King's industrial experience includes major design projects with Lawrence Livermore National Laboratory, as well as independent consulting projects. Prof. King has had a number of books published with titles including MATLAB, MathCAD, Exploring Engineering, and Engineering and Society.

David C. Kuncicky is a native Floridian. He earned his Baccalaureate in psychology, Master's in computer science, and Ph.D. in computer science from Florida State University. He has served as a faculty member in the Department  of Electrical Engineering at the FAMU–FSU College of Engineering and the Department of Computer Science at Florida State University. He has taught computer science and computer engineering courses for over 15 years. He has published research in the areas of intelligent hybrid systems and neural networks. He is currently the Director of Engineering at Bioreason, Inc. in Sante Fe, New Mexico.

Acknowledgments: Thanks to Steffie and Helen for putting up with my late nights and long weekends at the computer. Finally, thanks to Susan Bassett for having faith in my abilities, and for providing continued tutelage and support.

 **Ron Larsen** is a Professor of Chemical Engineering at Montana State University, and received his Ph.D. from the Pennsylvania State University. He was initially attracted to engineering by the challenges the profession offers, but also appreciates that engineering is a serving profession. Some of the greatest challenges he has faced while teaching have involved non-traditional teaching methods, including evening courses for practicing engineers and teaching through an interpreter at the Mongolian National University. These experiences have provided tremendous opportunities to learn new ways to communicate technical material. Dr. Larsen views modern software as one of the new tools that will radically alter the way engineers work, and his book *Introduction to MathCAD* was written to help young engineers prepare to meet the challenges of an ever-changing workplace.

Acknowledgments: To my students at Montana State University who have endured the rough drafts and typos, and who still allow me to experiment with their classes—my sincere thanks.

 **Sanford Leestma** is a Professor of Mathematics and Computer Science at Calvin College, and received his Ph.D. from New Mexico State University. He has been the long-time co-author of successful textbooks on Fortran, Pascal, and data structures in Pascal. His current research interest are in the areas of algorithms and numerical computation.

 **Jack Leifer** is an Assistant Professor in the Department of Mechanical Engineering at the University of Kentucky Extended Campus Program in Paducah, and was previously with the Department of Mathematical Sciences and Engineering at the University of South Carolina–Aiken. He received his Ph.D. in Mechanical Engineering from the University of Texas at Austin in December 1995. His current research interests include the analysis of ultra-light and inflatable (Gossamer) space structures.

Acknowledgments: I'd like to thank my colleagues at USC–Aiken, especially Professors Mike May and Laurene Fausett, for their encouragement and feedback; and my parents, Felice and Morton Leifer, for being there and providing support (as always) as I completed this book.

 **Richard M. Lueptow** is the Charles Deering McCormick Professor of Teaching Excellence and Associate Professor of Mechanical Engineering at Northwestern University. He is a native of Wisconsin and received his doctorate from the Massachusetts Institute of Technology in 1986. He teaches design, fluid mechanics, an spectral analysis techniques. Rich has an active research program on rotating filtration, Taylor Couette flow, granular flow, fire suppression, and acoustics. He has five patents and over 40 refereed journal and proceedings papers along with many other articles, abstracts, and presentations.

Acknowledgments: Thanks to my talented and hard-working co-authors as well as the many colleagues and students who took the tutorial for a "test drive." Special thanks to Mike Minbiole for his major contributions to Graphics Concepts with SolidWorks. Thanks also to Northwestern University for the time to work on a book. Most of all, thanks to my loving wife, Maiya, and my children, Hannah and Kyle, for supporting me in this endeavor. (Photo courtesy of Evanston Photographic Studios, Inc.)

 **Larry Nyhoff** is a Professor of Mathematics and Computer Science at Calvin College. After doing bachelor's work at Calvin, and Master's work at Michigan, he received a Ph.D. from Michigan State and also did graduate work in computer science at Western Michigan. Dr. Nyhoff has taught at Calvin for the past 34 years—mathematics at first and computer science for the past several years.

Acknowledgments: We thank our families—Shar, Jeff, Dawn, Rebecca, Megan, Sara, Greg, Julie, Joshua, Derek, Tom, Joan; Marge, Michelle, Sandy, Lory, Michael—for being patient and understanding. We thank God for allowing us to write this text.

Paul Riley is an author, instructor, and designer specializing in graphics and design for multimedia. He is a founding partner of CAD Support Associates, a contract service and professional training organization for computer-aided design. His 15 years of business experience and  20 years of teaching experience are supported by degrees in education and computer science. Paul has taught AutoCAD at the University of Massachusetts at Lowell and is presently teaching AutoCAD at Mt. Ida College in Newton, Massachusetts. He has developed a program, Computer-aided Design for Professionals that is highly regarded by corporate clients and has been an ongoing success since 1982.

Robert Rizza is an Assistant Professor of Mechanical Engineering at North Dakota State University, where he teaches courses in mechanics and computer-aided design. A native of Chicago, he received the Ph.D. degree from the  Illinois Institute of Technology. He is also the author of *Getting Started with Pro/ENGINEER*. Dr. Rizza has worked on a diverse range of engineering projects including projects from the railroad, bioengineering, and aerospace industries. His current research interests include the fracture of composite materials, repair of cracked aircraft components, and loosening of prostheses.

Peter Schiavone is a professor and student advisor in the Department of Mechanical Engineering at the University of Alberta, Canada. He received his Ph.D. from the University of Strathclyde, U.K. in 1988. He has authored  several books in the area of student academic success as well as numerous papers in international scientific research journals. Dr. Schiavone has worked in private industry in several different areas of engineering including aerospace and systems engineering. He founded the first Mathematics Resource Center at the University of Alberta, a unit designed specifically to teach new students the necessary *survival skills* in mathematics and the physical sciences required for success in first-year engineering. This led to the Students' Union Gold Key Award for outstanding contributions to the university. Dr. Schiavone lectures regularly to freshman engineering students and to new engineering professors on engineering success, in particular about maximizing students' academic performance.

Acknowledgements: Thanks to Richard Felder for being such an inspiration; to my wife Linda for sharing my dreams and believing in me; and to Francesca and Antonio for putting up with Dad when working on the text.

David I. Schneider holds an A.B. degree from Oberlin College and a Ph.D. degree in Mathematics from MIT. He has taught for 34 years, primarily at the University of Maryland. Dr. Schneider  has authored 28 books, with one-half of them computer programming books. He has developed three customized software packages that are supplied as supplements to over 55 mathematics textbooks. His involvement with computers dates back to 1962, when he programmed a special purpose computer at MIT's Lincoln Laboratory to correct errors in a communications system.

David I. Schwartz is an Assistant Professor in the Computer Science Department at Cornell University and earned his B.S., M.S., and Ph.D. degrees in Civil Engineering from State University of New York at Buffalo. Throughout his  graduate studies, Schwartz combined principles of computer science to applications of civil engineering. He became interested in helping students learn how to apply software tools for solving a variety of engineering problems. He teaches his students to learn incrementally and practice frequently to gain the maturity to tackle other subjects. In his spare time, Schwartz plays drums in a variety of bands.

Acknowledgments: I dedicate my books to my family, friends, and students who all helped in so many ways.

Many thanks go to the schools of Civil Engineering and Engineering & Applied Science at State University of New York at Buffalo where I originally developed and tested my UNIX and Maple books. I greatly appreciate the opportunity to explore my goals and all the help from everyone at the Computer Science Department at Cornell.

John T. Sears received the Ph.D. degree from Princeton University. Currently, he is a Professor and the head of the Department of Chemical Engineering at Montana State University. After leaving Princeton he worked in research at Brookhaven National Laboratory and Esso Research and Engineering, until he took a position at West Virginia University. He came to MSU in 1982, where he has served as the Director of the College of Engineering Minority Program and Interim Director for BioFilm Engineering. Prof. Sears has written a book on air pollution and economic development, and over 45 articles in engineering and engineering education.

Michael T. Snyder is President of Internet startup Appointments123.com. He is a native of Chicago, and he received his Bachelor of Science degree in Mechanical Engineering from the University of Notre Dame. Mike also graduated with honors from Northwestern University's Kellogg Graduate School of Management in 1999 with his Masters of Management degree. Before Appointments123.com, Mike was a mechanical engineer in new product development for Motorola Cellular and Acco Office Products. He has received four patents for his mechanical design work. "Pro/ENGINEER was an invaluable design tool for me, and I am glad to help students learn the basics of Pro/ENGINEER."

Acknowledgments: Thanks to Rich Lueptow and Jim Steger for inviting me to be a part of this great project. Of course, thanks to my wife Gretchen for her support in my various projects.

Jim Steger is currently Chief Technical Officer and cofounder of an Internet applications company. He graduated with a Bachelor of Science degree in Mechanical Engineering from Northwestern University. His prior work included mechanical engineering assignments at Motorola and Acco Brands. At Motorola, Jim worked on part design for two-way radios and was one of the lead mechanical engineers on a cellular phone product line. At Acco Brands, Jim was the sole engineer on numerous office product designs. His Worx stapler has won design awards in the United States and in Europe. Jim has been a Pro/ENGINEER user for over six years.

Acknowledgments: Many thanks to my co-authors, especially Rich Lueptow for his leadership on this project. I would also like to thank my family for their continuous support.

Royce Wilkinson received his undergraduate degree in chemistry from Rose-Hulman Institute of Technology in 1991 and the Ph.D. degree in chemistry from Montana State University in 1998 with research in natural product isolation from fungi. He currently resides in Bozeman, MT and is involved in HIV drug research. His research interests center on biological molecules and their interactions in the search for pharmaceutical advances.

Reviewers

We would like to thank everyone who has reviewed texts in this series.

ESource Reviewers

Christopher Rowe, *Vanderbilt University*
Steve Yurgartis, *Clarkson University*
Heidi A. Diefes-Dux, *Purdue University*
Howard Silver, *Fairleigh Dickenson University*
Jean C. Malzahn Kampe, *Virginia Polytechnic Institute and State University*
Malcolm Heimer, *Florida International University*
Stanley Reeves, *Auburn University*
John Demel, *Ohio State University*
Shahnam Navee, *Georgia Southern University*
Heshem Shaalem, *Georgia Southern University*
Terry L. Kohutek, *Texas A & M University*
Liz Rozell, *Bakersfield College*
Mary C. Lynch, *University of Florida*
Ted Pawlicki, *University of Rochester*
James N. Jensen, *SUNY at Buffalo*
Tom Horton, *University of Virginia*
Eileen Young, *Bristol Community College*
James D. Nelson, *Louisiana Tech University*
Jerry Dunn, *Texas Tech University*
Howard M. Fulmer, *Villanova UniversityBerkeley*
Naeem Abdurrahman *University of Texas, Austin*
Stephen Allan *Utah State University*
Anil Bajaj *Purdue University*
Grant Baker *University of Alaska–Anchorage*
William Beckwith *Clemson University*
Haym Benaroya *Rutgers University*
John Biddle *California State Polytechnic University*
Tom Bledsaw *ITT Technical Institute*
Fred Boadu *Duke University*
Tom Bryson *University of Missouri, Rolla*
Ramzi Bualuan *University of Notre Dame*
Dan Budny *Purdue University*
Betty Burr *University of Houston*
Dale Calkins *University of Washington*
Harish Cherukuri *University of North Carolina –Charlotte*
Arthur Clausing *University of Illinois*

Barry Crittendon *Virginia Polytechnic and State University*
James Devine *University of South Florida*
Ron Eaglin *University of Central Florida*
Dale Elifrits *University of Missouri, Rolla*
Patrick Fitzhorn *Colorado State University*
Susan Freeman *Northeastern University*
Frank Gerlitz *Washtenaw College*
Frank Gerlitz *Washtenaw Community College*
John Glover *University of Houston*
John Graham *University of North Carolina–Charlotte*
Ashish Gupta *SUNY at Buffalo*
Otto Gygax *Oregon State University*
Malcom Heimer *Florida International University*
Donald Herling *Oregon State University*
Thomas Hill *SUNY at Buffalo*
A.S. Hodel *Auburn University*
James N. Jensen *SUNY at Buffalo*
Vern Johnson *University of Arizona*
Autar Kaw *University of South Florida*
Kathleen Kitto *Western Washington University*
Kenneth Klika *University of Akron*
Terry L. Kohutek *Texas A&M University*
Melvin J. Maron *University of Louisville*
Robert Montgomery *Purdue University*
Mark Nagurka *Marquette University*
Romarathnam Narasimhan *University of Miami*
Soronadi Nnaji *Florida A&M University*
Sheila O'Connor *Wichita State University*
Michael Peshkin *Northwestern University*
Dr. John Ray *University of Memphis*
Larry Richards *University of Virginia*
Marc H. Richman *Brown University*
Randy Shih *Oregon Institute of Technology*
Avi Singhal *Arizona State University*
Tim Sykes *Houston Community College*
Neil R. Thompson *University of Waterloo*
Dr. Raman Menon Unnikrishnan *Rochester Institute of Technology*
Michael S. Wells *Tennessee Tech University*
Joseph Wujek *University of California, Berkeley*
Edward Young *University of South Carolina*
Garry Young *Oklahoma State University*
Mandochehr Zoghi *University of Dayton*

Contents

1

An Introduction to Engineering Problem Solving

GRAND CHALLENGE: WEATHER PREDICTION

Weather satellites provide a great deal of information to meteorologists, who attempt to predict the weather. Large volumes of historical weather data can also be analyzed and used to test models for predicting weather. In general, meteorologists can do a reasonably good job of predicting overall weather patterns. However, local weather phenomena, such as tornadoes, water spouts, and microbursts, are still very difficult to predict. Even predicting heavy rainfall or large hail from thunderstorms is often difficult. Although Doppler radar is useful in locating regions within storms that could contain tornadoes or microbursts, the radar detects the events as they occur and thus allows little time for issuing appropriate warnings to populated areas or aircraft passing through the region. Accurate and timely prediction of weather and associated weather phenomena is still an elusive goal.

OBJECTIVES

After reading this chapter, you should

- be acquainted with the fundamental problems in science and engineering in today's society
- understand the relationship of MATLAB to computer hardware and software, and
- understand a process for solving engineering problems.

1.1 GRAND CHALLENGES

Engineers solve real-world problems using scientific principles from disciplines that include computer science, mathematics, physics, and chemistry. It is this variety of subjects, and the challenge of solving real problems, that makes engineering so interesting and so rewarding. In this section, we present a group of **grand challenges**—fundamental problems in science and engineering with broad potential impact. The grand challenges were identified by the Office of Science and Technology Policy, in Washington, DC, as part of a research and development strategy for high-performance computing. The next set of paragraphs briefly presents some of these grand challenges and outlines the types of benefits that will come with their solutions; additional discussion of the individual challenges is presented at the beginning of each chapter. Just as the computer played an important part in the engineering achievements of the last 35 years, it will play an even greater role in solving problems related to these grand challenges.

The **prediction of change in weather**, **climate**, and **the global environment** requires that we understand the coupled atmosphere and ocean biosphere system. This includes understanding CO_2 dynamics in the atmosphere and ocean, ozone depletion, and climatological changes that are due to the release of chemicals or energy. These complex processes also include solar interactions. A major eruption from a solar storm near a "coronal hole" (a venting point for the solar wind) can eject vast amounts of hot gases from the sun's surface toward the earth's surface at speeds of over a million miles per hour. This ejection of hot gases bombards the earth with X rays and can interfere with communication and cause fluctuations in power lines. Learning to predict changes in weather, climate, and the global environment involves collecting large amounts of data for study and developing new mathematical models that can represent the interdependency of many variables.

Computerized speech understanding could revolutionize our communication systems, but many problems are involved with its development and implementation. It is currently possible to teach a computer to understand words from a small vocabulary spoken by one person. However, it is much more difficult to develop systems that are speaker independent and that understand words from large vocabularies and from different languages. In addition, subtle changes in one's voice, such as those caused by a cold or stress, can affect the performance of speech recognition systems. Even assuming that the computer can recognize the words, it may not be simple for the computer to determine their meaning. Many words are context dependent and thus cannot be analyzed independently. Intonation, such as raising one's voice, can change a statement into a question. While there are still many difficult problems to be addressed in the field of automatic speech recognition and understanding, exciting applications are everywhere. For example, imagine a telephone system that determines the languages being spoken over its lines and translates the speech signals so that each person hears the conversation in his or her native language.

The goals of the **Human Genome Project** are to locate, identify, and determine the functions of each of the 30,000+ human genes that are contained in the primary human genetic material, DNA or deoxyribonucleic acid. The majority of the human genome was sequenced in the year 2000 and 90 percent of the sequence of the genome's three billion base-pairs was published February 2001.

Now that most of the human genome has been mapped, the mission of researchers includes studies aimed at understanding how the human genome functions in the role of creating gene products, most notably the many proteins for which genes code.

The deciphering of the human genetic code and the determination of the function of each gene may lead to many technical advances, including the ability to detect, treat, and prevent many of the over 4,000 known human genetic diseases such as sickle-cell anemia and cystic fibrosis.

Substantial **improvements in vehicle performance** require more complex physical modeling in the areas of fluid dynamic behavior for three-dimensional flow fields and flow inside engine turbomachinery and ducts. Turbulence in fluid flows impacts the stability and control, thermal characteristics, and fuel performance of aerospace vehicles; modeling of this flow is necessary for the analysis of new configurations. The analysis of the aeroelastic behavior of vehicles also affects the development of new designs. The efficiency of combustion systems is related to vehicle performance as well, because attaining significant improvements in combustion efficiency requires an understanding of the relationships between the flows of the various substances and the chemistry that causes the substances to react. Vehicle performance is also being addressed through the use of onboard computers and microprocessors. For example, transportation systems in which small video screens are mounted on the dashboards of cars are currently being studied. The driver enters the destination into the computer, and the video screen shows the path, including street names, to get from the current location to the desired location. A communication network keeps the car's computer aware of any traffic jams along the path, so that it can automatically reroute the car if necessary. Other research on transportation addresses totally automated driving, with computers and networks handling all of the control and information interchange.

Enhanced oil and gas recovery will allow us to locate the estimated 300 billion barrels of oil reserves in the United States. Current techniques for identifying structures likely to contain oil and gas use seismic technology that can evaluate structures down to 20,000 feet below the surface. These techniques use a group of sensors, called a *sensor array*, that is located near the area to be tested. A ground shock signal is sent into the earth and is then reflected by the different geological layer boundaries and received by the sensors. Using sophisticated signal processing, the layer boundaries can be mapped, and some estimate can be made as to the materials in the various layers, such as sandstone, shale, and water. The ground shock signals can be generated in several ways: A hole can be drilled, and an explosive charge can be made in the hole; an explosive charge can be made on the surface; or a special truck that uses a hydraulic hammer can be used to pound the earth several times per second. Continued research is needed to improve the resolution of the information and to find methods of production and recovery that are economical and ecologically sound.

These grand challenges are only a few of the many interesting problems waiting to be solved by engineers and scientists. The solutions to problems of this magnitude will be the result of organized approaches that combine ideas and technologies. The use of computers and engineering problem-solving techniques will be a key element in the solution process.

1.2 COMPUTING SYSTEMS

Before we begin discussing MATLAB, we provide a brief discussion on computing, which is especially useful for those who have not had prior experience with computers. A **computer** is a machine designed to perform operations that are specified with a set of instructions, called a **program**. Computer **hardware** refers to the computer equipment, such as the keyboard, the mouse, the terminal, the hard disk, and the printer. Computer **software** refers to the programs that describe the steps we want the computer to perform.

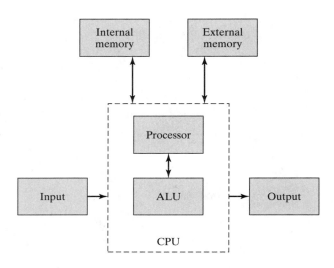

Figure 1.1. Internal organization of a computer.

1.2.1 Computer Hardware

All computers have a common internal organization, as shown in Figure 1.1. The **processor** is the part of the computer that controls all the other parts. It accepts input values (from a device such as a keyboard) and stores them in the computer's **memory**. It also interprets the instructions in a computer program. If we want to add two values, the processor will retrieve the values from memory and send them to the **arithmetic logic unit**, or ALU. The ALU performs the addition, and the processor then stores the result in memory. The processing unit and the ALU use internal memory composed of read-only memory (ROM) and random access memory (RAM) in their processing. Most data are stored in external memory or secondary memory using hard disk drives or floppy disk drives that are attached to the processor. The processor and ALU together are called the **central processing unit**, or CPU. A **microprocessor** is a CPU, which is contained in a single integrated circuit chip that contains millions of components in an area smaller than a postage stamp.

We usually instruct a computer to print the values that it has computed on the terminal screen or on paper, using a printer. Dot matrix printers use a matrix (or grid) of pins to produce the shape of a character on paper, whereas a laser printer uses a light beam to transfer images to paper. The computer can also write information to diskettes, which store the information magnetically. A printed copy of information is called a **hard copy**, and a magnetic copy of information is called an **electronic copy** or a soft copy.

Computers come in all sizes, shapes, and forms. (See Figure 1.2.) Personal computers (**PCs**) are small, inexpensive computers that are commonly used in offices, homes, and laboratories. PCs are also referred to as **microcomputers**. Their design is built around a microprocessor, such as the Pentium microprocessor, which can process millions of instructions per second (mips). Minicomputers are more powerful than microcomputers. Mainframes are even more powerful computers that are often used in businesses and research laboratories. A **workstation** is a minicomputer or mainframe computer that is small enough to fit on the top of a desk. **Supercomputers** are the fastest of all computers and can process billions of instructions per second. Because of their speed, supercomputers are capable of solving very complex problems that cannot feasibly be solved on other computers. Mainframes and supercomputers require special facilities and a specialized staff to run and maintain the computer systems.

Courtesy of Johnson Space Center

Courtesy of Getty Images, Inc.

Courtesy of Apple Computer, Inc.

Courtesy of Getty Images/EyeWire, Inc.

Courtesy of Boeing Commercial Airplane Group

Courtesy of PhotoEdit

Figure 1.2. A variety of computers for a variety of uses.

The type of computer needed to solve a particular problem depends on the requirements of the problem. If the computer is part of a home security system, a microprocessor is sufficient; if the computer is running a military grade flight simulator, a mainframe is probably needed. Computer **networks** allow computers to communicate with each other, so that they can share resources and information. For example, Ethernet is a commonly used local area network (LAN).

1.2.2 Computer Software

Computer software contains the instructions or commands that we want the computer to perform. There are several important categories of software, including operating systems, software tools, and language compilers. Figure 1.3 illustrates the interactions among these categories of software and the computer hardware. We now discuss each of these software categories in more detail.

Operating systems The program that controls or "operates" a computer is called the operating system. The operating system manages the computer's hardware such as the disk drives, terminal, keyboard, and modem. Other programs that want to access the computer's hardware must pass their requests to the operating system.

The operating system also manages the programs that are running on the computer. Modern computers may have tens or hundreds of programs running at any one time. The operating system is the traffic cop that schedules one program at a time to have access to the CPU.

Part of the operating system, called the kernel, is loaded when the computer is turned on. The kernel remains running until the computer is turned off. Common modern operating systems are Microsoft® Windows 2000, Microsoft® Windows XP, Linux, UNIX, and Apple Mac OS.

Operating systems also contain a group of programs called **utilities** that allow you to perform functions such as printing files, copying files from one disk to another, and listing the files that you have saved on a disk. Although these utilities are common to most operating systems, the commands themselves vary from operating system to operating system. For example, to list your files using DOS (a disk operating system used mainly with PCs), the command is *dir*; to list your files using UNIX (a powerful operating system frequently used with workstations) or linux (a powerful free operating system that is ported to many different hardware platforms), the command is *ls*. Some operating systems are referred to as *user-friendly*, because they simplify the interface with the user. Examples of user-friendly operating systems are the Macintosh environment and the Windows environment.

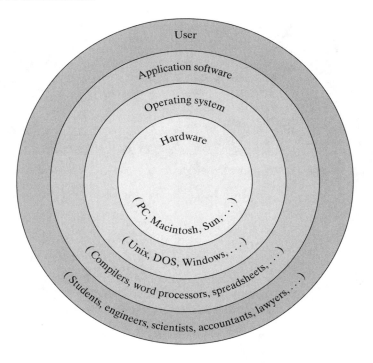

Figure 1.3. Interactions between software and hardware.

Because MATLAB programs can be run on many different platforms or hardware systems and because an individual computer can use different operating systems, it is not feasible to discuss the wide variety of operating systems that you might use while taking this course. We assume that your professor will provide you with information on the specific operating system that you need to use the computers available at your university. This information is also contained in the user's manual for the operating system.

Software Tools   Software tools are programs that have been written to perform common operations. For example, **word processors**, such as Microsoft® Word and WordPerfect®, are programs that have been written to help you enter and format text. Word processors allow you to move sentences and paragraphs and often have capabilities that enable you to enter mathematical equations and to check your spelling and grammar. Word processors are also used to enter computer programs and store them in files. Very sophisticated word processors allow you to produce well-designed pages that combine elaborate charts and graphics with text and headlines. These word processors use a technology called **desktop publishing**, which combines a very powerful word processor with a high-quality printer to produce professional-looking documents.

Spreadsheet programs are software tools that allow you to work easily with data that can be displayed in a grid of rows and columns. Spreadsheets were initially used for financial and accounting applications, but many science and engineering problems can be solved using spreadsheets as well. Most spreadsheet packages include plotting capabilities, so they can be especially useful in analyzing and displaying information. LOTUS 1-2-3 and Microsoft® Excel are popular spreadsheet programs.

Another popular group of software tools are database management programs such as Microsoft® Access, Microsoft® SQL Server, and ORACLE®. These programs allow you to store a large amount of data and then easily retrieve pieces of the data and format them into reports. Databases are used by large organizations, such as banks, hospitals, universities, hotels, and airlines, to store and organize crucial information. Databases are also used to analyze large amounts of scientific data. Meteorology and oceanography are examples of scientific fields that commonly require large databases for storage and analysis of data.

Computer-aided design (CAD) packages, such as AutoCAD®, ProE®, and Unigraphics®, allow you to define objects and then manipulate them graphically. For example, you can define an object and then view it from different angles or observe a rotation of the object from one position to another. CAD packages are frequently used in engineering applications.

MATLAB®, Mathematica®, Mathcad®, and Maple® are very powerful **mathematical computation** tools. Not only do these tools enable very powerful mathematical commands, but they also provide extensive capabilities for generating graphs. This combination of computational power and visualization power make them particularly useful tools for engineers.

If an engineering problem can be solved using a software tool, it is usually more efficient to use the software tool than to write a program in a computer language to solve the problem. However, many problems cannot be solved using software tools, or a software tool may not be available on the computer system that must be used for solving the problem. Thus, we also need to know how to write programs using computer languages. The distinction between a software tool and a computer language is becoming less clear as some of the more powerful tools, such as MATLAB and Mathematica, include their own languages in addition to specialized operations.

Computer Languages   A computer programming language is a notational form for relating instructions to a computer. Computer languages can be described in terms of levels. Low-level languages, or machine languages, are the most primitive languages. **Machine language** is tied closely to the design of the computer hardware. Because computer designs are based on two-state technology (i.e., computers are devices with two states, such as open or closed circuits, on or off switches, or positive or negative charges), machine language is written using two symbols, which are usually represented by the digits 0 and 1. Therefore, machine language is a binary language, and the instructions are written as sequences of 0s and 1s, called *binary strings*. Because machine language is closely tied to the design of the computer hardware, the machine language for a Sun Microsystems, Inc. computer is different from the machine language for a Silicon Graphics, Inc. computer.

An **assembly language** is a means of programming symbolically in machine language. Each line of code usually produces a single machine instruction. Assembly language is closely tied to the architecture of a specific processor such as the Intel Corporation 80×86 series or the Sun Microsystems, Inc., SPARC series. Programming in assembly language is certainly easier than programming in binary language, but it is still a tedious process.

EXAMPLE 1.1:

The assembly code listed here demonstrates typical assembly-language syntax. Each instruction is listed on a separate line and consists of an operation, or **opcode**, followed by its operands.

```
mov     cx,bx
shl     cx,8
shl     bx,6
add     bx,cx
add     ax,bx
mov     cx,es: [ax]
```

High-level languages are computer languages that have English-like commands and instructions and include languages such as C, Fortran, Ada, Pascal, COBOL, and BASIC. Writing programs in high-level languages is certainly easier than writing programs in machine language or in assembly language. However, a high-level language contains a large number of commands and an extensive set of **syntax** (or grammar) rules for using the commands. To illustrate the syntax and punctuation required by both software tools and high-level languages, we compute the area of a circle with a specified diameter in Table 1-1 using several different languages and tools. Notice both the similarities and the differences in this simple computation. Although we have included C as a high-level language, many people like to describe C as a midlevel language, because it allows access to low-level routines and is often used to define programs that are converted to assembly language.

Languages are also defined in terms of **generations**. The first generation of computer languages is machine language, the second generation is assembly language, and the third generation is high-level language. Fourth-generation languages, also referred to as **4GLs**, have not been developed yet and are described only in terms of characteristics and programmer productivity. The fifth generation of languages is called *natural languages*. To program in a fifth-generation language, one would use the syntax of natural speech. Clearly, the implementation of a natural language would require the achievement of one of the grand challenges: computerized speech understanding.

TABLE 1-1 Comparison of Software Statements.

SOFTWARE	EXAMPLE STATEMENT
MATLAB	`area = pi*((diameter/2)^2);`
C	`area = 3.141593*(diameter/2)*(diameter/2);`
Fortran	`area = 3.141593*(diameter/2.0)**2`
Ada	`area: = 3.141593*(diameter/2)**2;`
Pascal	`area: = 3.141593*(diameter/2)*(diameter/2)`
BASIC	`let a = 3.141593*(d/2)*(d/2)`
COBOL	`compute area = 3.141593*(diameter/2)*(diameter/2).`

Fortran (FORmula TRANslation) was developed in the mid-1950s for solving engineering and scientific problems. New standards updated the language over the years, and the current standard, Fortran 90, contains strong numerical computation capabilities, along with many of the new features and structures in languages such as C. **COBOL** (COmmon Business-Oriented Language) was developed in the late 1950s to solve business problems. Many legacy COBOL programs exist today and were a common source of the Year 2000 (Y2K) programming bug. **BASIC** (Beginner's All-purpose Symbolic Instruction Code) was developed in the mid-1960s and was used as an educational tool; in the 1980s, a BASIC interpreter was often included with the system software for a PC. **Pascal** was developed in the early 1970s and during the 1980s was widely used in computer science programs to introduce students to computing. **Ada** was developed at the initiative of the U.S. Department of Defense with the purpose of developing a high-level language appropriate to embedded computer systems that are typically implemented using microprocessors. The final design of the language was accepted in 1979. The language was named in honor of Ada Lovelace, who developed instructions for doing computations on an analytical machine in the early 1800s. **C** is a general-purpose language that evolved from two languages, BCPL and B, that were developed at Bell Laboratories, Inc. in the late 1960s. In 1972, Dennis Ritchie developed and implemented the first C compiler on a DEC PDP–11 computer at Bell Laboratories, Inc. The language was originally developed in order to write the UNIX operating system. Until that time, most operating systems were written in assembly language. C became very popular for system development because it was hardware independent (unlike assembly code). Because of its popularity in both industry and in academia, it became clear that a standard definition of it was needed. A committee of the American National Standards Institute (ANSI) was created in 1983 to provide a machine-independent and unambiguous definition of C. In 1989, the C ANSI standard was approved. **C++** is an object-oriented programming language that is a superset of the C language. Much of the early development of C++ was made in the mid-1980s by Bjarne Stroustrup at Bell Laboratories, Inc. The major features that C++ adds to C are inheritance, abstract classes, overloaded operators, and a form of dynamic type binding (virtual functions). During the 1990s, C++ became the dominant programming language for applications in such diverse fields as engineering, finance, telecommunications, embedded systems, and computer-aided design. In 1997, the International Standards Organization (ISO) approved a standard for C++.

Executing a Computer Program   A program written in a high-level language, such as C, must be translated into machine language before the instructions can be executed by the computer. A special program called a **compiler** is used to perform this translation.

Thus, in order to be able to write and execute C programs on a computer, the computer's software must include a C compiler.

If any errors (often called **bugs**) are detected by the compiler during compilation, corresponding error messages are printed. We must correct our program statements and then perform the compilation step again. The errors identified during this stage are called **compile errors**, or **compile-time errors**. For example, if we want to divide the value stored in a variable called *sum* by 3, the correct expression in C is **sum/3**. If we incorrectly write the expression using the backslash, as in **sum\3**, we will have a compiler error. The process of compiling, correcting statements (or **debugging**), and recompiling must often be repeated several times before the program compiles without compiler errors. When there are no compiler errors, the compiler generates a program in machine language that performs the steps specified by the original C program. The original C program is referred to as the **source program**, and the machine-language version is called an **object program**. Thus, the source program and the object program specify the same steps, but the source program is specified in a high-level language, and the object program is specified in machine language.

Once the program has compiled correctly, additional steps are necessary to prepare the object program for **execution**. This preparation involves **linking** other machine-language statements to the object program and then **loading** the program into memory. After these linking and loading steps have been performed, the program's commands are then executed by the computer. New errors, called **execution errors**, **runtime errors**, **logic errors**, or **program bugs**, may be identified in this stage. Execution errors often cause the termination of a program. For example, the program statements may attempt to perform a division by zero, which generates an execution error. Some execution errors do not stop the program from executing, but they cause incorrect results to be computed. These types of errors can be caused by programmer errors in determining the correct steps in the solutions and by errors in the data processed by the program. When execution errors occur because of errors in the program statements, we must correct the errors in the source program and then begin again with the compilation step. Even when a program appears to execute properly, we must check the results carefully to be sure that they are correct. The computer will perform the steps precisely as we specify. If we specify the wrong steps, the computer will execute these wrong (but syntactically legal) steps and present us with an answer that is incorrect.

The processes of compilation, linking/loading, and execution are outlined in Figure 1.4. The process of converting an assembly language program to binary language is performed by an **assembler** program, and the corresponding processes are called assembly, linking/loading, and execution.

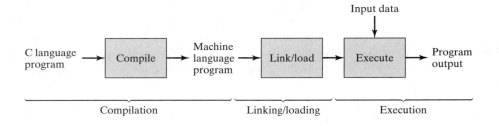

Figure 1.4. Program compilation/loading, linking, and execution.

Executing a MATLAB *Program*   In the MATLAB environment, we can develop and execute programs, or scripts, that contain MATLAB commands. We can also execute a MATLAB command, observe the results, and then execute another MATLAB command that interacts with the information in memory, observe its results, and so on. This **interactive environment** does not require the formal compilation, linking/loading, and execution process that we described for high-level computer languages. However, errors in the syntax of a MATLAB command are detected when the MATLAB environment attempts to translate the command, and logic errors can cause execution errors when the MATLAB environment attempts to execute the command.

Software Life Cycle   In 1955, the cost of a typical computer solution was estimated to be broken down as follows: 15% for the software development and 85% for the associated computer hardware. Over the years, the cost of hardware has dramatically decreased, while the cost of software has increased. In 1985, it was estimated that these numbers had essentially switched, with 85% of the cost being for the software and 15% for the hardware. With the majority of the cost of a computer solution residing in the development of software, a great deal of attention has been given to understanding the development of a software solution.

The development of a software project generally follows defined steps, or cycles, which are collectively called the **software life cycle**. These steps typically include project definition, detailed specification, coding and modular testing, integrated testing, and maintenance. Data indicate that the corresponding percentages of effort involved can be estimated as shown in Table 1-2. From these estimates, it is clear that software maintenance contributes a significant part of the cost of a software system. This maintenance includes adding enhancements to the software, fixing errors identified as the software is used, and adapting the software to work with new hardware and software. The ease of providing maintenance is directly related to the original definition and specification of the solution, because these steps lay the foundation for the rest of the project. The problem-solving process that we present in the next section emphasizes the need to define and specify the solution carefully before beginning to code or test it.

One of the techniques that has been successful in reducing the cost of software development, in terms of both time and money is the development of **software prototypes**. Instead of waiting until the software system has been developed to let users work with it, a prototype of the system is developed early in the life cycle. This prototype does not have all the functions required of the final software, but it allows the user to use it early in the life cycle and to make desired modifications to the specifications. Making changes earlier in the life cycle is both cost effective and time effective. Because of its powerful commands and its graphics capabilities, MATLAB is especially effective in developing software prototypes. Once the MATLAB prototype is correctly performing the desired operations and the users are happy with the user–software interaction, the final solution may be the MATLAB program, or the final solution may be converted to another language for implementation with a specific computer or piece of instrumentation.

TABLE 1-2 Software Life Cycle Phases.

LIFE CYCLE	PERCENT OF EFFORT
Definition	3%
Specification	15%
Coding and Modular Testing	14%
Integrated Testing	8%
Maintenance	60%

As an engineer, it is very likely that you will need to modify or add additional capabilities to existing software. These modifications will be much simpler if the existing software is well structured and readable and if the documentation that accompanies the software is up-to-date and clearly written. Even with powerful tools such as MATLAB, it is important to write well-structured and readable code. For these reasons, we stress the importance of developing good habits that make software more readable and self-documenting.

1.3   AN ENGINEERING PROBLEM-SOLVING METHODOLOGY

Problem solving is a key part not only of engineering courses, but also of courses in computer science, mathematics, physics, and chemistry. Therefore, it is important to have a consistent approach to solving problems. It is also helpful if the approach is general enough to work for all these different areas, so that we do not have to learn one technique for solving mathematics problems, a different technique for solving physics problems, and so on. The problem-solving technique that we present works for engineering problems and can be tailored to solve problems in other areas as well. However, it does assume that we are using a computer to help solve the problem.

The process, or methodology, for problem solving that we will use throughout this text has **five steps**:

1.   State the problem clearly.
2.   Describe the input and output information.
3.   Work the problem by hand (or with a calculator) for a simple set of data.
4.   Develop a MATLAB solution.
5.   Test the solution with a variety of data.

We now discuss each of these steps, using data collected from a physics laboratory experiment as an example. Assume that we have collected a set of temperatures from a sensor on a piece of equipment that is being used in an experiment. The temperature measurements are taken every 30 seconds, for 5 minutes, during the experiment. We want to compute the average temperature, and we also want to plot the temperature values.

1.3.1   Problem Statement

The first step is to state the problem clearly. It is extremely important to give a clear, concise statement of the problem, in order to avoid any misunderstandings. For this example, the statement of the problem is as follows:

Compute the average of a set of temperatures. Then plot the time and temperature values.

1.3.2   Input/Output Description

The second step is to describe carefully the information that is given to solve the problem and then to identify the values to be computed. These items represent the input and the output for the problem and collectively can be called **input/output**, or **I/O**. For many problems, it is useful to create a diagram that shows the input and output. At this point, the program is called an **abstraction** because we are not defining the steps to determine the output; instead, we are only showing the information that is used to compute the output. The **I/O diagram** for this example is as follows:

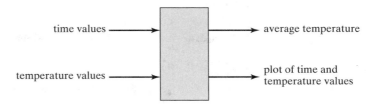

1.3.3 Hand Example

The third step is to work the problem by hand or with a calculator, using a simple set of data. This step is very important and should not be skipped, even for simple problems. This is the step in which you work out the details of the solution to the problem. If you cannot take a simple set of numbers and compute the output (either by hand or with a calculator), you are not ready to move on to the next step. You should reread the problem and perhaps consult reference material. For this problem, the only calculation is computing the average of a set of temperature values. Assume that we use the following data for the hand example:

TIME (MINUTES)	TEMPERATURE (DEGREES F)
0.0	105
0.5	126
1.0	119

By hand, we compute the average to be $(105 + 126 + 119)/3$, or 116.6667 degrees F.

1.3.4 MATLAB Solution

Once you can work the problem for a simple set of data, you are ready to develop an **algorithm**, which is a step-by-step outline of the solution to the problem. For simple problems such as this one, the algorithm can be written immediately using MATLAB commands. For more complicated problems, it may be necessary to write an outline of the steps and then decompose the steps into smaller steps that can be translated into MATLAB commands. One of the strengths of MATLAB is that its commands match very closely to the steps that we use to solve engineering problems. Thus, the process of determining the steps to solve the problem also determines the MATLAB commands. Observe that the MATLAB steps match closely to the solution steps from the hand example:

```
%    Compute average temperature and
%    plot the temperature data.
%
time = [0.0, 0.5, 1.0];
temps = [105, 126, 119];
average = mean(temps)
plot (time,temps),title('Temperature Measurements'),
xlabel ('Time, minutes'),
ylabel ('Temperature, degrees F'),grid
```

The words that follow percent signs are comments to help us in reading the MATLAB statements. If a MATLAB statement assigns or computes a value, it will also print the value on the screen if the statement does not end in a semicolon. Thus, the values of **time** and **temps** will not be printed, because the statements that assign them values end with semicolons. The value of the average will be computed and printed on the screen, because the statement that computes it does not end with a semicolon. Finally, a plot of the time and temperature data will be generated.

1.3.5 Testing

The final step in our problem-solving process is testing the solution. We should first test the solution with the data from the hand example, because we have already computed the solution to it. When the previous statements are executed, the computer displays the following output:

```
average =
   116.6667
```

A plot of the data points is also shown on the screen. Because the value of the average computed by the program matches the value from the hand example, we now replace the data from the hand example with the data from the physics experiment, yielding the following program:

```
%  Compute average temperature and
%  plot the temperature data.
%
time = [0.0, 0.5, 1.0, 1.5, 2.0, 2.5, 3.0, . . .
          3.5, 4.0, 4.5, 5.0];
temps = [105, 126, 119, 129, 132, 128, 131, . . .
          135, 136, 132, 137];
average = mean(temps)
plot (time,temps),title('Temperature Measurements'),
xlabel ('Time, minutes'),
ylabel ('Temperature, degrees F'),grid
```

When these commands are executed, the computer displays the following output:

```
average =
   128.1818
```

The plot in Figure 1.5 is also shown on the screen.

SUMMARY

A set of grand challenges was presented to illustrate some of the exciting and difficult problems that currently face engineers and scientists. Because the solutions to most engineering problems, including the grand challenges, involve the use of computers, we also presented a summary of the components of a computer system, from computer hardware to computer software. We then introduced a five-step problem-solving methodology that we will use to develop computer solutions to problems. The five steps are as follows:

    a.   State the problem clearly.
    b.   Describe the input and output information.
    c.   Work the problem by hand (or with a calculator) for a simple set of data.
    d.   Develop a MATLAB solution.
    e.   Test the solution with a variety of data.

This process will be used throughout the text as we develop solutions to problems.

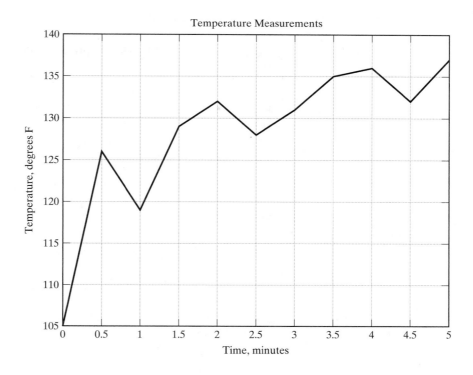

Figure 1.5.   Temperatures collected in the physics experiment.

KEY TERMS

algorithm	hard copy	soft copy
assembly language	hardware	software
bugs	high-level languages	software life cycle
central processing unit	logic errors	software prototypes
compile errors	machine language	syntax
compiler	microprocessor	utilities
debugging	opcode	workstation
electronic copy	operating system	

Problems

1.   Match the following terms:

    ____CPU   A. low-level language

    ____kernel   B. operating system

    ____MATLAB   C. steps for solving a problem

    ____assembler   D. hardware

    ____algorithm   E. high-level language

    ____Java   F. mathematical computation tool

2. Create an I/O diagram for the following description of a problem:

   Each monitoring station in a group of five stations produces representative quality assurance ratings for that station. Compute the mean quality assurance rating for each of the five monitoring stations. Present the results in a bar graph. Then find the station with the maximum rating and compute the mean of the five station means.

3. Solve the previous problem by hand for the following data:

STATION 1	STATION 2	STATION 3	STATION 4	STATION 4
10.6	5.6	1.3	6.5	3.4
9.8	7.2	1.5	6.2	
4.5	3.4	1.8		
	5.8	2.1		
	5.9	1.2		
		1.1		

4. Write an algorithm that describes the steps that you used to solve the previous problem. Write the algorithm in plain English.

2

MATLAB Environment

GRAND CHALLENGE: VEHICLE PERFORMANCE

Wind tunnels are test chambers built to generate precise wind speeds. Accurate scale models of new aircraft can be mounted on force-measuring supports in the test chamber, and then measurements of the forces acting on the models can be made at many different wind speeds and angles of the models relative to the wind direction. Some wind tunnels can operate at hypersonic velocities, generating wind speeds of thousands of miles per hour. The sizes of wind tunnel test sections vary from a few inches across to sizes large enough to accommodate a business jet. At the completion of a wind tunnel test series, many sets of data have been collected that can be used to determine the lift, drag, and other aerodynamic performance characteristics of a new aircraft at its various operating speeds and positions.

OBJECTIVES

After reading this chapter, you should

- understand the MATLAB screen layout, windows, and interactive environment
- be able to create, initialize, and graph scalars, vectors, and matrices
- be able to perform arithmetic operations on scalars and arrays
- be acquainted with special MATLAB constants and matrices
- be able to format and label graphs, and
- be able to apply the principles of this chapter to a problem in velocity computation.

2.1 CHARACTERISTICS OF THE MATLAB ENVIRONMENT

The MATLAB software was originally developed to be a **matrix laboratory**, that is software for manipulation of matrices. Today's MATLAB has capabilities far beyond those of the original MATLAB and is an interactive system and programming language for general scientific and technical computation. Its basic element is a matrix, which we discuss in detail in the next section. Because the MATLAB commands are similar to the way in which we express engineering steps in mathematics, writing computer solutions in MATLAB is much quicker than writing computer solutions using a high-level language, such as C or Fortran. In this section, we explain the differences between the student version and the professional version of MATLAB, and we give you some initial MATLAB workspace information.

2.1.1 Student Edition Version 6

MATLAB provides an inexpensive student version of Releases 12 and 13 for the Microsoft Windows and the Linux operating systems. The student version of MATLAB includes the following components of the professional version:

- the basic MATLAB engine and development environment
- the MATLAB notebook
- SIMULINK
- some of the Symbolic Math Toolbox functions

Matrix sizes are unlimited in the student version. The amount of memory on your computer is the limiting factor for maximum matrix size. This is the same for the professional version of MATLAB. The SIMULINK toolbox for the student version is limited to 300 modeling blocks. The student version's CD contains the full electronic documentation.

Other toolboxes must be purchased separately. Not all toolboxes are available for the student version. For more information, please see the MATLAB website.

The student version of MATLAB has a different prompt. Instead of **>>**, the prompt is **EDU>>**.

We assume that MATLAB is already installed on the computer that you are using. (If you have purchased *The Student Edition of MATLAB*, follow the installation instructions in the manual that accompanies the software.) The discussions and programs developed in this text will run properly using either the student edition or the professional version. We will assume that the input interaction uses a keyboard and a mouse.

2.1.2 MATLAB Windows

To begin MATLAB, use your mouse to click on the MATLAB icon, which should be located on the desktop or in a menu bar. If you are using a UNIX operating system, type **matlab** at the shell prompt. You should see the MATLAB **prompt** (**>>** or **EDU>>**), which tells you that MATLAB is waiting for you to enter a command. To exit MATLAB, type **quit** or **exit** at the MATLAB prompt.

MATLAB uses several display windows. Among these are a **command window**, a **graphics window**, and an **edit window**. As you execute commands, the appropriate windows will automatically appear. You can choose the window that you want to be active by simply placing the mouse icon within the window and clicking the mouse button.

When you first enter MATLAB, the command window will be the active window. The command window is used to enter commands and data and to print results.

EXAMPLE 2.1:   For example, to compute the square root of 9, type the following command:

```
sqrt (9)
```

The following output will be displayed:

```
ans =
      3
```

The graphics window is used to display plots and graphs. For example, to see the graphics window, type

```
plot([1,2,4,9,16],[1,2,3,4,5])
```

at the prompt. The graphics window will automatically appear, and a plot of the two vectors will be displayed. If you want to see some of the advanced graphics capabilities of MATLAB, as well as see a demonstration of the graphics window, enter **demo** at the prompt. This initiates the MATLAB Demo Window, a graphical demonstration environment that illustrates some of the different types of operations that can be performed with MATLAB.

The edit window is used to create and modify M-files, which are files that contain a program or script of MATLAB commands. The edit window appears when program files are created or loaded. To view the M-File Editor/Debugger window, type

```
edit
```

at the prompt.

2.1.3 Managing the Environment

The MATLAB run-time environment consists of the variables that are placed in memory during a single session. To see the current run-time environment, type **who** or **whos** at the prompt.

EXAMPLE 2.2:   For example, enter the following commands to create and initialize the scalar **X** and the vector **Y**:

```
X = 5;
Y = [4, 5];
```

Then type **whos** to see the environment. The following should be displayed:

```
Name        Size          Bytes  Class
X           1x1               8  double array
Y           1x2              16  double array

Grand total is 3 elements using 24 bytes
```

An alternative way to view the environment is to choose **View** and then **Show Workspace** from the menu bar. The command **clear** removes all variables from memory. Type **clear** and then **whos** to see that the environment is cleared:

```
clear
whos
```

There are several commands for clearing windows. The **clc** command clears the command window, and the **clf** command clears the current figure and thus clears the graph window. In general, it is a good idea to start programs with the **clear** and **clf** commands, to be sure that the memory has been cleared and that the graph window has been cleared and reset.

To preserve the contents of the workspace environment between sessions, you must save the environment to a file. The default format is a binary file called a MAT-file. To save the workspace to a file named **myspace.mat**, type

```
save myspace
```

at the prompt or choose **File** and then **Save Workspace As** from the menu bar. To restore a workspace from a disk file, use the **load** command.

```
load myspace
```

The commands listed in Table 2.1 are helpful for locating and manipulating MATLAB disk files. You can run external commands from within the MATLAB environment by using the exclamation-point character (**!**). The particular external commands will vary, depending on the operating system that you are using. For example, in the Windows/DOS environment, the **ver** command shows the version of the operating system:

```
!ver
```

The results of executing the **ver** command on your operating system will be displayed. For example, you might see something like

```
Microsoft Windows XP [Version 5.1.2600]
```

The result shows that the author's computer is running the Microsoft Windows XP operating system.

You can use the shell escape to execute external editors. This capability will be useful when you are writing MATLAB programs and wish to use your favorite editor.

It is important to know how to abort a command in MATLAB. For example, there may be times when your commands cause the computer to print seemingly endless lists of numbers or when the computer goes into an infinite loop. In these cases, hold down the CTRL key and press the C key to generate a local abort within MATLAB.

2.1.4 Getting Help

The Student Edition of MATLAB includes extensive on-line help tools. There are three ways to get on-line help from within MATLAB: a command-line help function (**help**), a separate windowed help function (**helpwin**), and an HTML-based documentation set.

TABLE 2.1 MATLAB Disk Operations.

MATLAB COMMAND	DESCRIPTION	DOS EQUIVALENT	UNIX EQUIVALENT
dir	list directory	DIR	ls
delete	delete file	DEL	rm
cd	change directory	CD	cd
path	show current path	PATH	printenv

To use the command-line help function, type **help** in the command window:

```
help
```

A list of help topics will appear, shown as follows:

```
HELP topics:
matlab\general     - General-purpose commands.
matlab\ops         - Operators and special characters.
matlab\lang        - Programming language constructs.
matlab\elmat       - Elementary matrices and matrix
                     manipulation.
matlab\elfun       - Elementary math functions.
matlab\specfun     - Specialized math functions.
etc., etc.
```

To get help on a particular topic, type **help topic**.

EXAMPLE 2.3:   For example, to get help on the tangent function, type **help tan**:

```
help tan
```

The following should be displayed:

```
TAN      Tangent.

   TAN (X) is the tangent of the elements of X.
   See also ATAN, ATAN2.
```

To use the windowed help screen, select **Help** and then **Help Matlab** from the menu bar. A windowed version of the help list will appear. If you are connected to the World Wide Web, you can subscribe to the MathWorks Newsletter and the MathWorks MATLAB Digest. These are both e-mail lists that will keep you up to date on the latest MATLAB improvements and technical suggestions. These and other services can be accessed at the MathWorks website: www.mathworks.com.

2.2   SCALARS, VECTORS, AND MATRICES

When solving engineering problems, it is important to be able to visualize the data related to the problem. Sometimes the data is just a single number, such as the radius of a circle. Other times, the data may be a coordinate on a plane that can be represented as a pair of numbers, with one number representing the x coordinate and the other number representing the y coordinate. In another problem, we might have a set of four x–y–z coordinates that represent the four vertices of a pyramid with a triangular base in a three-dimensional space. We can represent all of these examples using a special type of data structure called a **matrix**. A matrix is a set of numbers arranged in a rectangular grid of rows and columns. Thus, a single point can be considered a matrix with one row and one column, an x–y coordinate can be considered a matrix with one row and two columns, and a set of four x–y–z coordinates can be considered a matrix with four rows and three columns. Some examples of such matrices are as follows:

$$\mathbf{A} = [3.5] \quad \mathbf{B} = [1.5 \ \ 3.1]$$

$$\mathbf{C} = \begin{bmatrix} -1 & 0 & 0 \\ 1 & 1 & 0 \\ 1 & -1 & 0 \\ 0 & 0 & 2 \end{bmatrix}$$

Note that the data within a matrix are written within brackets. When a matrix has one row and one column, we can also refer to the number as a **scalar**. Similarly, when a matrix has one row or one column, we can refer to it as a **row vector** or a **column vector**, respectively.

When we use a matrix, we need a way to refer to individual elements or numbers in the matrix. A simple method for specifying an element in the matrix uses the row and column number. For example, if we refer to the value in row 4 and column 3 in the matrix $\mathbf{C}$ in the previous example, there is no ambiguity: We are referring to the value 2. We specify the row and column number as subscripts of the name of the matrix; thus, $\mathbf{C}_{4,3}$ is equal to 2. To refer to the entire matrix, we use the name of the matrix without subscripts, as in $\mathbf{C}$, or we use brackets around the reference to an individual element that uses letters instead of numbers for the subscripts, as in $[\mathbf{C}_{i,j}]$. In formal mathematical notation, the name of a matrix is usually an uppercase letter, but appears in lowercase when used with subscripted references, as in $\mathbf{C}$, $\mathbf{c}_{4,3}$ and $[\mathbf{c}_{i,j}]$. However, because MATLAB is case sensitive, $\mathbf{C}$ and $\mathbf{c}$ represent different matrices. Therefore, in your MATLAB programs, you will need to consistently use either uppercase or lowercase letters when referring to a specific matrix.

The size of a matrix is specified by the number of rows and columns it contains. Thus, using our previous example, $\mathbf{C}$ is a matrix with four rows and three columns, or a 4×3 matrix. If a matrix contains m rows and n columns, then the number of values is the product of m and n; thus, $\mathbf{C}$ contains 12 values. If a matrix has the same number of rows as columns, it is called a **square matrix**.

PRACTICE!

Answer the accompanying questions about the following matrix:

$$\mathbf{G} = \begin{bmatrix} 0.6 & 1.5 & 2.3 & -0.5 \\ 8.2 & 0.5 & -0.1 & -2.0 \\ 5.7 & 8.2 & 9 & 1.5 \\ 0.5 & 0.5 & 2.4 & 0.5 \\ 1.2 & -2.3 & -4.5 & 0.5 \end{bmatrix}$$

1. What is the size of $\mathbf{G}$?
2. Give the subscript references for all locations that contain the value 0.5.

In MATLAB programs, we assign names to the scalars, vectors, and matrices that we use. The following rules apply to these variable names:

- Variable names must start with a letter.

- Other than the first letter, variable names can contain letters, digits, and the underscore character (_). To test whether a name is a legitimate variable name, use the **isvarname** command. The answer 1 means true, and 0 means false. For example,

```
isvarname Vector
ans =
      1
```

- Variable names can be any length, but only the first N characters are used by MATLAB. The value of N varies, depending on the version of MATLAB that you are using. For Version 6 Release 13, the value of N is 63. You can see the value of N on your system by typing **namelengthmax**.
- Variables cannot have the same name as any of MATLAB's keywords. To see a list of all MATLAB keywords, type **iskeyword**.
- MATLAB allows you to use the names of its built-in functions as variable names. This is a dangerous practice, since you can overwrite the meaning of a function such as **sin**. To check whether a name is a built-in function, use the **which** command. For example,

```
which sin
sin is a built-in function.
```

Because MATLAB is case sensitive, the names Time, TIME, and time all represent different variables.

2.2.1   Initializing Variables

We present four methods for initializing matrices in MATLAB. The first method explicitly lists the values, the second reads data from a file, the third uses the colon operator, and the fourth reads data from the keyboard.

Explicit Lists   The simplest way to define a matrix is to use a list of numbers, as shown in the following example, which defines the matrices **A**, **B**, and **C** that we used in our previous example:

```
A = [3.5];
B = [1.5,3.1];
C = [-1,0,0; 1,1,0; 1,-1,0; 0,0,2];
```

These statements are examples of the assignment statement, which consists of a variable name followed by an equals sign and the data values to assign to the variable. The data values are enclosed in brackets in row order. Semicolons separate the rows, and the values in the rows can be separated by commas or blank spaces. A value can contain a plus or minus sign and a decimal point, but it cannot contain a comma, as in 32,010.

When we define a matrix, MATLAB will also print the value of the matrix on the screen unless we suppress the printing with a semicolon after the definition. In our examples, we will generally include the semicolon to **suppress printing**. However, when you are first learning to define matrices, it is helpful to see their values. Therefore, you may want to omit the semicolon after a matrix definition until you are confident that you know how to define matrices properly. The **who** and **whos** commands are also very helpful. The **who** command lists the matrices that you have defined, and the **whos** command lists the matrices and their sizes.

A matrix can also be defined by listing each row on a separate line, as in the following set of MATLAB commands:

```
C = [-1,   0, 0
      1,   1, 0
      1,  -1, 0
      0,   0, 2];
```

If there are too many numbers in a row of the matrix to fit on one line, you can continue the statement on the next line, but a comma and three periods (an **ellipsis**) are needed at the end of the line to indicate that the row is to be continued. For example, if we want to define a row vector **F** with 10 values, we could use either of the following statements:

```
F = [1,52,64,197,42,-42,55,82,22,109];
F = [1, 52, 64, 197, 42, -42, ...
     55, 82, 22, 109];
```

MATLAB also allows you to define a matrix using another matrix that has already been defined. For example, consider the following statements:

```
B = [1.5, 3.1];
S = [3.0 B];
```

These commands are equivalent to the following:

```
B = [1.5, 3.1];
S = [3.0 1.5 3.1];
```

We can also change values in a matrix or include additional values by using a reference to specific locations. Thus, the command

```
S(2) = -1.0;
```

changes the second value in the matrix **S** from 1.5 to −1.0.

We can also extend a matrix by defining new elements. If we execute the command

```
S(4) = 5.5;
```

the matrix **S** will have four values instead of three. If we execute the command

```
S(8) = 9.5;
```

then the matrix **S** will have eight values, and the values of **S(5)**, **S(6)**, and **S(7)** are automatically set to zero, because no values were given for them.

We can also change a range of values:

```
S(2:4) = [3.9 6.9 0.4];
```

In addition, we can change a range of values to a single value:

```
S(5:7) = 1;
```

Finally, we can change an arbitrary subset of a matrix at once by making a list of indices:

```
I = [1 3 7];
S(I) = 42;
S(I + 1) = [73 24 16];
```

PRACTICE!

Give the size of each matrix. Then check your answers by entering the appropriate commands in MATLAB. In these problems, a matrix definition may refer to a **previously defined matrix.**

```
a.  B = [2; 4; 6; 10]
b.  C = [5 3 5; 6 2 -3]
c.  E = [3 5 10 0; 0 0 . . .
        0 3; 3 9 9 8]
d.  T = [4 24 9]
    Q = [T 0 T]
e.  V = [C(2,1); B]
f.  A(2,1) = -3
```

Data Files   Matrices can also be initialized from information that has been stored in a data file. MATLAB can interface with two different types of data files: MAT-files and ASCII files. The MAT-file, mentioned in the previous section, contains data stored in a memory-efficient binary format, whereas an ASCII file contains information stored in ASCII characters. MAT-files are preferable for storing data that are going to be generated and used only by MATLAB programs, because its storage is more efficient than that of ASCII files.

ASCII files are useful if the data are to be shared (imported or exported) with programs other than MATLAB programs. An ASCII data file that is going to be used with a MATLAB program should contain only numeric information, and each row of the file should contain the same number of data values. The file can be generated using your favorite word processor. It can also be generated by running a program written in a computer language, such as a C program. In addition, the file can be generated by a MATLAB program using one of the following forms of the **save** command, where **<fname>** is a filename and **<vlist>** is a list of variables to be saved:

```
save <fname> <vlist> -ascii  Use 8-digit text format.
save <fname> <vlist> - ascii - double  Use 16-digit text
format.
save <fname> <vlist> -ascii-double-tabs  Delimit elements
with tabs.
```

EXAMPLE 2.4:

For example, to create the matrix **Z** and save it to the file **data_2.dat** in eight-digit text format, use the following commands:

```
Z = [5 3 5; 6 2 3];
save data_2.dat Z -ascii;
```

This command causes each row of the matrix **Z** to be written to a separate line in the data file. The **.mat** extension is not added to an ASCII file. However, as we illustrated in this example, we recommend that ASCII filenames include the extension **.dat**, so that it is easy to distinguish them from MAT-files and M-files. If you view the file **data_2.dat** with a word processor such as WordPad or vi, it should appear as follows:

```
5.0000000e+000 3.0000000e+000 5.0000000e+000
6.0000000e+000 2.0000000e+000 3.0000000e+000
```

Suppose that an ASCII file named **data_3.dat** contains a set of values that represent the time and corresponding distance of a runner from the starting line in a race. Each time and its corresponding distance value are on a separate line of the data file. Thus, the first few lines in the data file might have the following form:

```
0.0 0.0
0.1 3.5
0.2 6.8
```

Entering the **load** command followed by the filename at the prompt will cause the information to be read into a matrix with the same name as the data file. If the **Current Directory** child window is available, all saved files in the current directory are visible and can be double clicked on to start the import wizard, to help import the data.

EXAMPLE 2.5:

For example, the following statements will load data from the file **data_3.dat** and store the data in the matrix **data_3**, which has three rows and two columns:

```
load data_3.dat
data_3
```

You can then type the name of the variable to view its contents. The following should be displayed:

```
data_3 =
     0          0
     0.1000     3.5000
     0.2000     6.8000
```

PRACTICE!

Create a file containing a matrix description, such as the following example matrix:

```
2.2 -23.1  14.5
3.0   45   -18.3
23.2 -34.4  16
```

Use a text editor—for example, WordPad or NotePad in the Windows environment. In the Unix environment, the available editors might be Emacs, vi, or pico. Be sure the file is saved in text or ASCII file format. Load the data into MATLAB using the **load** command. Type the variable name (same as the filename without the extension) to verify that the data were loaded correctly.

Colon Operator   The colon operator is a very powerful operator for creating new matrices. For example, the colon operator can be used to create vectors from a matrix. When a colon is used in a matrix reference in place of a specific subscript, the colon represents the entire row or column. For example, if we use the **data_3** matrix that was read from a data file in the previous discussion, the following commands will store the first column of **data_3** in the column vector **x** and the second column of **data_3** in the column vector **y**:

```
x = data_3(:,1);
y = data_3(:,2);
```

The colon operator can also be used to generate new matrices. When a colon is used to separate two integers, all of the integers between the two specified integers are generated. For example, the following notation generates a vector named **H** that contains the integers from 1 to 8:

```
H = 1:8;
```

When colons are used to separate three numbers, values between the first and third numbers are generated, using the second number as the increment. For example, the following notation generates a row vector named **time** that contains the numbers from 0.0 to 5.0 in increments of 0.5:

```
time = 0.0:0.5:5.0;
```

The increment can also be negative, as shown in the following example, which generates the numbers 10, 9, 8, . . . 0 in the row vector named **values**:

```
values = 10:-1:0;
```

Incrementing can be done even if the end value is not a multiple of the value of the increment:

```
odd = 0: 0.7: 2;
```

The colon operator can also be used to select a **submatrix** from another matrix. For example, consider the following matrix:

$$
C = \begin{bmatrix} -1 & 0 & 0 \\ 1 & 1 & 0 \\ 1 & -1 & 0 \\ 0 & 0 & 2 \end{bmatrix}
$$

If we execute the commands

```
C_partial_1 = C(:,2:3);
C_partial_2 = C(3:4,1:2);
```

we have defined the following matrices:

$$
\text{C\_partial\_1} = \begin{bmatrix} 0 & 0 \\ 1 & 0 \\ -1 & 0 \\ 0 & 2 \end{bmatrix} \qquad \text{C\_partial\_2} = \begin{bmatrix} 1 & -1 \\ 0 & 0 \end{bmatrix}
$$

If the colon notation defines a matrix with invalid subscripts, as in **C(5:6,:)**, an error message is displayed.

In MATLAB, it is valid to have a matrix that is empty. For example, the following statements will each generate an **empty matrix**:

```
a = [];
b = 4:-1:5;
```

Note that an empty matrix is different than a matrix that contains only zeros.

The use of the expression **C(:)** is equivalent to one long column matrix that contains the first column of **C**, followed by the second column of **C**, and so on.

An operator that is very useful with matrices is the **transpose** operator. The transpose of a matrix **A** is denoted by **A′** and represents a new matrix in which the rows of **A** are transformed into the columns of **A′**. For now we will use the transpose operator only to turn a row vector into a column vector and a column vector into a row vector. This operation can be very useful when printing vectors. For example, suppose that we generate two vectors **x** and **y**. We then want to print the values of the vectors such that **x(1)** and **y(1)** are on the same line, **x(2)** and **y(2)** are on the same line, and so on. A simple way to do this is as follows:

```
x = 0:4;
y = 5:5:25;
[x' y']
```

The output generated by these statements is

```
0 5
1 10
2 15
3 20
4 25
```

The transpose operator will also be useful in generating some of the tables specified in the problems at the end of this chapter.

PRACTICE!

Create the following matrix in MATLAB:

$$G = \begin{bmatrix} 0.6 & 1.5 & 2.3 & -0.5 \\ 8.2 & 0.5 & -0.1 & -2.0 \\ 5.7 & 8.2 & 9.0 & 1.5 \\ 0.5 & 0.5 & 2.4 & 0.5 \\ 1.2 & -2.3 & -4.5 & 0.5 \end{bmatrix}$$

Calculate the results of each of the following commands by hand, and then check your answers by typing each command into MATLAB:

```
a.   A = G(:,2)
b.   C = 10:15
c.   D = [4:9; 1:6]
d.   F = 0.0:0.1:1.0
e.   T1 = G(4:5,1:3)
f.   T2 = G(1:2:5,:)
```

User Input   The values of a matrix can also be entered through the keyboard by using the **input** command, which displays a text string and then waits for input. The value entered by the user is stored in the specified variable. If more than one value is to be entered by the user, the set of values must be enclosed in brackets. If the user strikes the Return key without entering input values, an empty matrix is returned. If the command does not end with a semicolon, the values entered for the matrix are printed.

Consider the following command:

```
z = input('Enter values for z in brackets:');
```

When this command is executed, the text string **Enter values for z in brackets** is displayed on the terminal screen. The user can then enter an expression, such as **[5.1 6.3 -18.0]**, which then specifies values for **z**. Because this **input** command ends with a semicolon, the values of **z** are not printed when the command is completed.

2.2.2 Printing Matrices

There are several ways to print the contents of a matrix. The simplest way is to enter the name of the matrix. The name of the matrix will be repeated, and the values of the matrix will be printed, starting on the next line. There are also several commands that can be used to print matrices that provide more control over the form of the output.

Display Format   When elements of a matrix are printed, integers are always printed as integers. Values with decimal fractions are printed using a **default format** (called a **short format**) that shows four decimal digits. MATLAB allows you to specify other formats (see Table 2.2) that show more significant digits. For example, to specify that we want values to be displayed in a decimal format with 14 decimal digits, we use the command **format long**. We can return the format to a decimal format with four decimal digits by using the command **format short**. Two decimal digits are displayed when the format is specified as **format bank**.

When a value is very large or very small, decimal notation does not work satisfactorily. For example, a value that is used frequently in chemistry is Avogadro's constant, whose value to four significant places is 602,300,000,000,000,000,000,000. Obviously, we need a more manageable notation for very large values, like Avogadro's constant, or for very small values, like 0.0000000031. **Scientific notation** expresses a value as a number between 1 and 10 multiplied by a power of 10. In scientific notation, Avogadro's constant becomes 6.023×10^{23}. This form is also referred to as a mantissa (6.023) and an exponent (23). In MATLAB, values in scientific notation are printed with the letter "e" to separate the mantissa and the exponent, as in 6.023e+23. If we want MATLAB to print values in scientific notation with 5 significant digits, we use the command **format short e**. To specify scientific notation with 16 significant digits, we use the command **format long e**. We can also enter values in a matrix using scientific notation, but it is important to omit blanks between the mantissa and the exponent, as MATLAB will interpret 6.023 e+23 as two values (6.023 and e+23), whereas 6.023e+23 will be interpreted as one value.

TABLE 2.2 Numeric Display Formats.

MATLAB COMMAND	DISPLAY	EXAMPLE
format short	4 decimal digits	15.2345
format long	14 decimals digits	15.23453333333333
format short e	4 decimals digits	1.5235e+01
format long e	15 decimals digits	1.523453333333333e+01
format bank	2 decimals digits	15.23
format +	+, −, blank	+

Another format command is **format +**. When a matrix is printed with this value, the only characters printed are plus and minus signs. If a value is positive, a plus sign will be printed; if a value is zero, a space will be skipped; if a value is negative, a minus sign will be printed. This format allows us to view a large matrix in terms of its signs. Otherwise, we would not be able to see it easily, because there might be too many values in a row to fit on a single line.

For long and short formats, a common scale factor is applied to the entire matrix if the elements become very large. This scale factor is printed along with the scaled values.

Finally, the command **format compact** suppresses many of the line feeds that appear between matrix displays and allows more lines of information to be seen together on the screen. In our example output, we will assume that this command has been executed. The command **format loose** will return to the less compact display mode.

Printing Text or Matrices   The **disp** command can be used to display text enclosed in single quote marks. It can also be used to print the contents of a matrix without printing the matrix's name. Thus, if a scalar **temp** contains a temperature value in degrees Fahrenheit, we could print the value on one line and the units on the next line by using the following commands:

```
disp(temp); disp('degrees F')
```

If the value of **temp** is 78, then the output will be the following:

```
78
degrees F
```

Note that the two **disp** commands were entered on the same line, so that they would be executed together. To display the values on the same line, you create one string from the two parts:

```
disp([num2str(temp) ' degrees F'])
```

Formatted Output   The **fprintf** command gives you even more control over the output than you have with the **disp** command. In addition to printing both text and matrix values, you can specify the format to be used in printing the values, and you can specify when to skip to a new line. If you are a C programmer, you will be familiar with the syntax of the command. With few exceptions, the MATLAB **fprintf** command uses the same formatting specifications as the C **fprintf()** function. We will not cover all of the format options in this text. Please refer to an ANSI C textbook for a complete description of the **fprintf()** options.

The general form of this command is

```
fprintf(format-string,var, . . .)
```

The format string contains format specifications to be printed. The string may contain a combination of literal text and format specifications that define how the data is to be displayed. The remainder of the arguments is a list of names of the matrices to be printed.

The format specifiers take the following form:

```
%[flags] [width] [.precision] type
```

Each specifier begins with the **%** character. The flags, width, and precision format fields are optional. Examples of **type fields** are

```
%e display in exponential notation
%f display in fixed-point or decimal notation
%g display using %e or %f, depending on which is shorter
```

Strings are displayed literally. The **%** character may be displayed by using **%%**. The tab and newline characters are displayed by using **\t** and **\n**, respectively.

EXAMPLE 2.6:

As an example, enter the following commands:

```
temp = 98.6;
fprintf('The temperature is %f degrees F.\n', temp);
```

The following should be displayed:

```
The temperature is 98.600000 degrees F.
```

The **width field** controls the minimum number of characters to be printed. It must be a nonnegative decimal integer. The **precision field** is preceded by a period (.) and specifies the number of decimal places after the decimal point for exponential and fixed-point types. Note in the preceding example that the default precision is six digits. The precision field has a different meaning for other types.

EXAMPLE 2.7:

For example, enter the following statement:

```
fprintf('The temperature is %3.1f degrees.\n', temp);
```

The following output will be produced:

```
The temperature is 98.6 degrees.
```

The **flags field** is used to designate the padding characters and to format numerical signs. The zero (0) flag with a fixed-point type will cause the leading digits to be padded with zeros instead of spaces. A minus flag (−) left-justifies the converted argument in its field. A plus flag (+) always prints a sign character (+ or −).

EXAMPLE 2.8:

For example, enter the following statement:

```
fprintf('The temperature is %08.1f degrees.\n', temp);
```

The following output will be produced:

```
The temperature is 000098.6 degrees.
```

The **fprintf** statement allows you to have a great deal of control over the form of the output. We will use it frequently in our examples to help you become familiar with it.

EXAMPLE 2.9:

Try this example to see the power of the **fprintf** command when it is used with a matrix. Enter these two statements:

```
X = 0:0.2:2;
fprintf(' %3.1f * 5 = %4.1f \n',X,X)
```

The following output will be displayed:

```
0.0 * 5 =   0.2
0.4 * 5 =   0.6
0.8 * 5 =   1.0
1.2 * 5 =   1.4
1.6 * 5 =   1.8
2.0 * 5 =   0.0
0.2 * 5 =   0.4
0.6 * 5 =   0.8
1.0 * 5 =   1.2
1.4 * 5 =   1.6
1.8 * 5 =   2.0
```

PRACTICE!

You may wonder how the **fprintf** command responds when displaying a nonscalar matrix. The answer is that the format string is cycled in column major order through the elements of the matrix until all the elements have been used. For example, type the following two statements:

```
A = [ 2,   4,   6
      3,   7,   8
      6,  28,  48 ];

fprintf('func(%1.0f,%2.0f) = %2.0f \n', A);
```

The following output will be displayed:

```
func(2,  3) =  6
func(4,  7) = 28
func(6,  8) = 48
```

Execute the following statement to initialize the matrix **S**:

```
S = [1, 2, 3, 4; 3, 7, 21, 14; 28, 14, 17, 7 ];
```

Practice by attempting to duplicate the following output using a **fprintf** statement:

```
Game 1 score: Us:  3 Them: 28
Game 2 score: Us:  7 Them: 14
Game 3 score: Us: 21 Them: 17
Game 4 score: Us: 14 Them:  7
```

2.2.3   *XY* Plots

In this section, we show you how to generate a simple *xy* plot from data stored in two vectors. Assume that we want to plot the data shown in the accompanying table, collected from an experiment with a remotely controlled model car. The experiment is repeated 10 times, and we have measured the distance that the car travels for each trial.

TRIAL	DISTANCE, FT
1	58.5
2	63.8
3	64.2
4	67.3
5	71.5
6	88.3
7	90.1
8	90.6
9	89.5
10	90.4

Assume that the time values are stored in a vector called **x** and that the distance values are stored in a vector called **y**. To plot these points, we use the **plot** command, with **x** and **y** as arguments:

```
plot(x,y)
```

The plot in Figure 2.1 is automatically generated. (Slight variations in the scaling of the plot may occur, depending on the computer type and the size of the graphics window.)

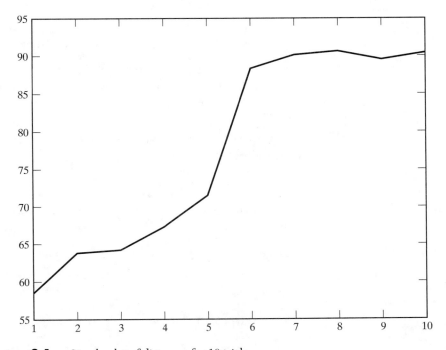

Figure 2.1.   Simple plot of distances for 10 trials.

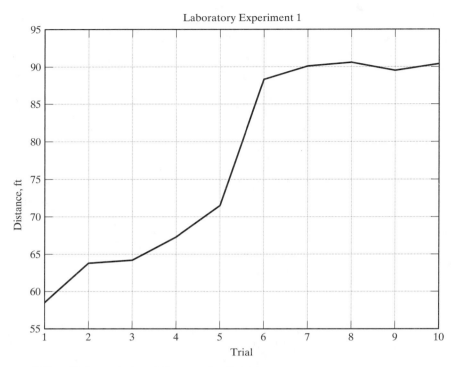

Figure 2.2. Enhanced plot of distances for 10 trials.

Good engineering practice requires that we include units and a title in our plot. Therefore, we include the following commands, which add a title, *x*- and *y*-axis labels, and a background grid:

```
plot(x,y),title('Laboratory Experiment 1'),
xlabel('Trial'),ylabel('Distance, ft'),grid
```

These commands generate the plot in Figure 2.2.

If you display a plot and then continue with more computations, MATLAB will generate and display the graph in the graphics window and then return immediately to execute the rest of the commands in the program. Because the plot window is replaced by the command window when MATLAB returns to finish the computations, you may want to use the **pause** command to halt the program temporarily, to give yourself a chance to study the plot. Execution will continue when any key is pressed. If you want to pause for a specified number of seconds, use the **pause(n)** command, which will cause an execution pause for **n** seconds before continuing. The **print** command will print the contents of the graphics window on the printer attached to the computer or to a file.

PRACTICE!

Practice plotting several trigonometric functions. Create a vector **X** to hold a series of *x*-axis values from 0 to 2 π.

```
X = 0.0:pi/100:2*pi;
```

Define vector **Y1** to be a function of vector **X**:

```
Y1 = cos(X*4);
```

Plot the function using the following statement:

```
plot (X,Y1);
```

By default, the execution of a second plot statement will erase the first plot. You can layer plots on top of one another by using the **hold on** statement. Execute the following statements to ensure that both functions are plotted on the same graph, as shown in Figure 2.3.

```
Y2 = sin (X);
hold on;
plot (X, Y2);
```

2.3 SCALAR AND ARRAY OPERATIONS

Arithmetic computations are specified by matrices and constants combined with arithmetic operations. In this section, we first discuss operations involving only scalars; then we extend the operations to include element-by-element operations.

2.3.1 Scalar Operations

The arithmetic operations between two scalars are shown in Table 2.3. Expressions containing scalars and scalar operations can be evaluated and stored in a specified variable, as in the following statement, which specifies that the values in **a** and **b** are to be added and the sum is to be stored in **x**:

```
x = a + b;
```

This assignment statement should be interpreted as specifying that the value in **a** is added to the value in **b**, and the sum is stored in **x**. If we interpret assignment statements in this way, we are not disturbed by the following valid MATLAB statement:

```
count = count + 1;
```

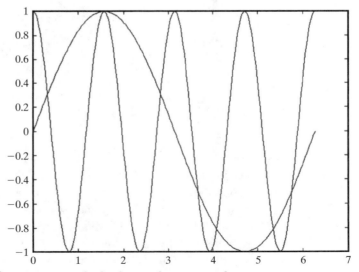

Figure 2.3. Creating multiple plots on the same graph.

TABLE 2.3 Arithmetic Operations between Two Scalars.

OPERATION	ALGEBRAIC FORM	MATLAB FORM
addition	$a + b$	`a + b`
subtraction	$a - b$	`a - b`
multiplication	$a \times b$	`a*b`
division	$\dfrac{a}{b}$	`a/b`
exponentiation	a^b	`a^b`

Clearly, this statement is not a valid algebraic statement, but in MATLAB it specifies that 1 is to be added to the value in **count**, and the result stored back in **count**. Therefore, it is equivalent to specifying that the value in **count** should be incremented by 1.

It is important to recognize that a variable can store only one value at a time. For example, suppose that the following MATLAB statements were executed one after another:

```
time = 0.0;
time = 5.0;
```

The value 0.0 is stored in the variable **time** when the first statement is executed and is then replaced by the value 5.0 when the second statement is executed.

When you enter an expression without specifying a variable to store the result, the result or answer is automatically stored in a variable named **ans**. Each time that a new value is stored in **ans**, the previous value is lost.

PRACTICE!

Demonstrate that the default variable **ans** exists by executing the following commands:

```
16/2^2;
whos
```

The following output will be displayed:

```
Name        Size            Bytes   Class
ans         1x1                 8   double array
Grand total is 1 element(s) using 8 bytes
```

ans can then be used like any other variable, but be aware that it will change when the next answer is generated.

2.3.2 Precedence of Arithmetic Operations

Because several operations can be combined in a single arithmetic expression, it is important to know the order in which operations are performed. Table 2.4 contains the precedence of arithmetic operations in MATLAB. Note that this precedence follows the rules of standard algebraic precedence.

Assume that we want to use integration to compute the area of a trapezoid from numerical integration where the base is horizontal and the two edges are vertical. The variable **base** contains the length of the base, and **height_1** and **height_2**

TABLE 2.4 Precedence of Arithmetic Operations.

PRECEDENCE	OPERATION
1	parentheses, innermost first
2	exponentiation, left to right
3	multiplication and division, left to right
4	addition and subtraction, left to right

contain the two heights. The area of the trapezoid can be computed using the following MATLAB statement:

```
area = 0.5*base*(height_1 + height_2);
```

Suppose that we omit the parentheses in the expression:

```
area = 0.5*base*height_1 + height_2;
```

The preceding statement would be executed as if it were the following statement:

```
area = (0.5*base*height_1) + height_2;
```

Note that although an incorrect answer has been computed, there are no error messages printed to alert us to the problem. Therefore, it is important to be very careful when converting equations into MATLAB statements. Adding extra parentheses is an easy way to ensure that computations are done in the order that you want.

If an expression is long, break it into multiple statements. For example, consider the following equation:

$$f = \frac{x^3 - 2x^2 + x - 6.3}{x^2 + 0.05005 - 3.14}$$

The value of f could be computed using the following MATLAB statements:

```
numerator = x^3 - 2*x^2 + x - 6.3;
denominator = x^2 + 0.05005*x - 3.14;
f = numerator/denominator;
```

It is better to use several statements that are easy to understand than to use one statement that requires careful thought to figure out the order of operations.

PRACTICE!

Give MATLAB commands to compute the values in each problem. Assume that the variables in the equations are scalars and that they have been assigned values.

1. Correction factor in pressure calculation:

$$\text{factor} = 1 + \frac{b}{v} + \frac{c}{v^2}$$

2. Slope between two points:

$$\text{slope} = \frac{y_2 - y_1}{x_2 - x_1}$$

3. Resistance of a parallel circuit:

$$\text{resistance} = \frac{1}{\dfrac{1}{r_1} + \dfrac{1}{r_2} + \dfrac{1}{r_3}}$$

4. Pressure loss from pipe friction:

$$\text{loss} = f \cdot p \cdot \frac{1}{d} \cdot \frac{v^2}{2}$$

2.3.3 Computational Limitations

The variables stored in a computer can assume a wide range of values. For most computers, the range of values extends from 10^{-308} to 10^{308}, which should be enough to accommodate most computations. However, it is possible for the result of an expression to be outside of this range. For example, suppose that we execute the following commands:

```
x = 2.5e200;
y = 1.0e200;
z = x*y;
```

If we assume the range of values is from 10^{-308} to 10^{308}, then the values of **x** and **y** are within the allowable range. However, the value of **z** is 2.5e400, and this value exceeds the range. This error is called **exponent overflow**, because the exponent of the result of an arithmetic operation is too large to store in the computer's memory. In MATLAB, the result of an exponent overflow is ∞.

Exponent underflow is a similar error, caused by the exponent of the result of an arithmetic operation being too small to store in the computer's memory. Using the same allowable range, we obtain an exponent underflow error with the following commands:

```
x = 2.5e-200;
y = 1.0e200;
z = x/y;
```

Again, the values of **x** and **y** are within the allowable range, but the value of **z** is 2.5e–400. Because the value of **z** is less than the minimum allowable value, we have caused an exponent underflow error to occur. In MATLAB, the result of an exponent underflow is zero.

We also know that **division by zero** is an invalid operation. If an expression results in a division by zero in MATLAB, the result of the division is ∞. MATLAB will print a warning message, and subsequent computations will continue.

2.3.4 Array Operations

An **array operation** is performed element by element. For example, suppose that **A** is a row vector with five elements and **B** is a row vector with five elements. One way to generate a new row vector **C** with values that are the products of corresponding values in **A** and **B** is the following:

```
C(1) = A(1)*B(1);
C(2) = A(2)*B(2);
```

```
C(3) = A(3)*B(3);
C(4) = A(4)*B(4);
C(5) = A(5)*B(5);
```

These commands are essentially scalar commands, because each command multiplies a single value by another single value and stores the product in a third value. To indicate that we want to perform an element-by-element multiplication between two matrices of the same size, we use an asterisk preceded by a period. Thus, the five preceding statements can be replaced by the following:

```
C = A.*B;
```

Omitting the period before the asterisk is a serious error, because the statement then specifies a matrix operation, not an element-by-element operation.

For addition and subtraction, array operations and matrix operations are the same, so we do not need to distinguish between them. However, array operations for multiplication, division, and exponentiation are different than matrix operations for multiplication, division, and exponentiation, so we need to include a period to specify an array operation. These rules are summarized in Table 2.5.

Element-by-element operations, or array operations, apply not only to operations between two matrices of the same size, but also to operations between a scalar and a nonscalar. However, multiplication of a matrix by a scalar can be written either way. Thus, the two following sets of statements are equivalent for a matrix **A**:

```
B = 3*A;
B = 3.*A;
C = A/5;
C = A./5;
```

The resulting matrices **B** and **C** will be the same size as **A**.

To illustrate the array operations for vectors, consider the following two row vectors:

$$A = \begin{bmatrix} 2 & 5 & 6 \end{bmatrix} \quad B = \begin{bmatrix} 2 & 3 & 5 \end{bmatrix}$$

If we compute the array product of **A** and **B** using the statement

```
C = A. *B;
```

then **C** will contain the following values:

$$C = \begin{bmatrix} 4 & 15 & 30 \end{bmatrix}$$

TABLE 2.5   Element-by-Element Operations.

OPERATION	ALGEBRAIC FORM	MATLAB FORM
addition	$a + b$	`a + b`
subtraction	$a - b$	`a - b`
multiplication	$a \times b$	`a.*b`
division	$\dfrac{a}{b}$	`a./b`
exponentiation	a^b	`a.^b`

The array division command, demonstrated in the statement

```
C = A./B;
```

will generate a new vector in which each element of **A** is divided by the corresponding element of **B**. Thus, **C** will contain the following values:

```
C = [1 1.6667 1.2]
```

Array exponentiation is also an element-by-element operation. For example, consider the following statement:

```
A = [2, 5, 6];
B = [2, 3, 5];
C = A.^2;
D = A.^B;
```

The vectors **C** and **D** are as follows:

```
C = [4 25 36]   D = [4 125 7776]
```

We can also use a scalar base to a vector exponent, as in

```
C = 3.0.^A;
```

which generates a vector with the following values:

```
C = [9 243 729]
```

This vector could also have been computed with the statement

```
C = (3).^A;
```

If you are not sure that you have written the correct expression, always test it with simple examples like the ones we have just used.

The previous examples used vectors, but the same rules apply to matrices with rows and columns, as shown by the following statements:

```
d = [1:5; -1:-1:-5];
p = d.*5;
q = d.^3;
```

The values of these matrices are shown as follows:

$$d = \begin{bmatrix} 1 & 2 & 3 & 4 & 5 \\ -1 & -2 & -3 & -4 & -5 \end{bmatrix}$$

$$p = \begin{bmatrix} 5 & 10 & 15 & 20 & 25 \\ -5 & -10 & -15 & -20 & -25 \end{bmatrix}$$

$$q = \begin{bmatrix} 1 & 8 & 27 & 64 & 125 \\ -1 & -8 & -27 & -64 & -125 \end{bmatrix}$$

PRACTICE!

Give the values in the vector **C** after execution of each statement, where **A** and **B** contain the following values:

```
A = [2 -1 5 0]   B = [3 2 -1 4]
```

Check your answers using MATLAB.

```
a.  C = B + A - 3;
b.  C = 2*A + A.^B;
c.  C = A./B;
d.  C = A.^B;
e.  C = 2.^B + A;
f.  C = 2*B/3.*A;
```

2.4 SPECIAL VALUES AND SPECIAL MATRICES

MATLAB includes a number of predefined constants, special values, and special matrices that are available to the programs we write. Most of these special values and special matrices are generated by MATLAB using functions. A MATLAB function typically uses inputs called **arguments** to compute a matrix, although some functions do not require any input arguments. In this section, we give some examples of MATLAB functions.

2.4.1 Special Values

The scalar values that are available for use in MATLAB programs are described in the following list:

pi Represents π.

i, j Represents $\sqrt{-1}$.

Inf Represents infinity, which typically occurs as a result of a division by zero. A warning message will be printed when this value is computed; if you display a matrix containing this value, the value will print as **Inf**.

NaN Represents Not-a-Number and typically occurs when an expression is undefined, as in the division of zero by zero.

clock Represents the current time in a six-element row vector containing year, month, day, hour, minute, and seconds.

date Represents the current date in a character string format, such as **02-Jan-2001**.

eps Represents the epsilon floating-point precision for the computer being used. This epsilon precision is the smallest amount with which two values can differ in the computer.

ans Represents a value computed by an expression, but not stored in a variable name.

2.4.2 Special Matrices

MATLAB contains a group of functions that generate special matrices; we present some of these functions here.

Matrix of Zeros The **zeros** function generates a matrix containing all zeros. If the argument to the function is a scalar, as in **zeros(6)**, the function will generate a square matrix using the argument as both the number of rows and the number of columns. If the function has two scalar arguments, as in **zeros(m, n)**, the function will generate a matrix with m rows and n columns. Because the **size** function returns two scalar arguments that represent the number of rows and the number of columns in a matrix, we can use the **size** function to generate a matrix of zeros that is the same size as another matrix. The following statements illustrate these various cases:

```
A = zeros(3);
B = zeros(3,2);
C = [1 2 3 ; 4 2 5];
D = zeros(size(C));
```

The matrices generated are as follows:

$$A = \begin{bmatrix} 0 & 0 & 0 \\ 0 & 0 & 0 \\ 0 & 0 & 0 \end{bmatrix} \quad B = \begin{bmatrix} 0 & 0 \\ 0 & 0 \\ 0 & 0 \end{bmatrix}$$

$$C = \begin{bmatrix} 1 & 2 & 3 \\ 4 & 2 & 5 \end{bmatrix} \quad D = \begin{bmatrix} 0 & 0 & 0 \\ 0 & 0 & 0 \end{bmatrix}$$

Matrix of Ones The **ones** function generates a matrix containing all ones, just as the **zeros** function generates a matrix containing all zeros. If the argument to the function is a scalar, as in **ones(6)**, the function will generate a square matrix using the argument as both the number of rows and the number of columns. If the function has two scalar arguments, as in **ones(m, n)**, the function will generate a matrix with m rows and n columns. To generate a matrix of ones that is the same size as another matrix, use the **size** function to determine the correct number of rows and columns. The following statements illustrate these various cases:

```
A = ones(3);
B = ones(3,2);
C = [1 2 3 ; 4 2 5];
D = ones(size(C));
```

The matrices generated are as follows:

$$A = \begin{bmatrix} 1 & 1 & 1 \\ 1 & 1 & 1 \\ 1 & 1 & 1 \end{bmatrix} \quad B = \begin{bmatrix} 1 & 1 \\ 1 & 1 \\ 1 & 1 \end{bmatrix}$$

$$C = \begin{bmatrix} 1 & 2 & 3 \\ 4 & 2 & 5 \end{bmatrix} \quad D = \begin{bmatrix} 1 & 1 & 1 \\ 1 & 1 & 1 \end{bmatrix}$$

Identity Matrix An identity matrix is a matrix with ones on the main diagonal and zeros elsewhere. For example, the following matrix is an identity matrix with four rows and four columns:

$$\begin{bmatrix} 1 & 0 & 0 & 0 \\ 0 & 1 & 0 & 0 \\ 0 & 0 & 1 & 0 \\ 0 & 0 & 0 & 1 \end{bmatrix}$$

Note that the **main diagonal** is the diagonal containing elements in which the row number is the same as the column number. Therefore, the subscripts for elements on the main diagonal are (1,1), (2,2), (3,3), and so on.

In MATLAB, identity matrices can be generated using the **eye** function. The arguments of the **eye** function are similar to those for the **zeros** function and the **ones** function. If the function has one scalar argument, as in **eye(6)**, the function will generate an identity matrix using the argument as both the number of rows and the number of columns. If the function has two scalar arguments, as in **eye(m, n)**, the function will generate an identity matrix with *m* rows and *n* columns. To generate an identity matrix that is the same size as another matrix, use the **size** function to determine the correct number of rows and columns. Although most applications use a square identity matrix, the definition can be extended to nonsquare matrices. The following statements illustrate these various cases:

```
A = eye(3);
B = eye(3,2);
C = [ 1 2 3 ; 4 2 5];
D = eye(size(C));
```

The matrices generated are as follows:

$$A = \begin{bmatrix} 1 & 0 & 0 \\ 0 & 1 & 0 \\ 0 & 0 & 1 \end{bmatrix} \quad B = \begin{bmatrix} 1 & 0 \\ 0 & 1 \\ 0 & 0 \end{bmatrix}$$

$$C = \begin{bmatrix} 1 & 2 & 3 \\ 4 & 2 & 5 \end{bmatrix} \quad D = \begin{bmatrix} 1 & 0 & 0 \\ 0 & 1 & 0 \end{bmatrix}$$

We recommend that you do not name an identity matrix **i**, because **i** will then not represent $\sqrt{-1}$ in any statements that follow.

Diagonal Matrices   The **diag** function can be used to create a diagonal matrix or extract one of the diagonals of a matrix. To extract the main diagonal from the foregoing matrix **A**, type

```
diag(A)
```

The following column vector will be generated:

```
ans =
     1
     1
     1
```

Other diagonals can be extracted by passing a second parameter k to **diag.** The second parameter denotes the position of the diagonal from the main diagonal ($k = 0$). Using the example matrix **C** from before, type

```
diag(C,1)
```

The following column vector will be generated:

```
ans =
        2
        5
```

If the first argument to **diag** is a vector **V**, then this function generates a square matrix. If $k = 0$, then the elements of **V** are placed on the main diagonal. If $k > 0$, they are placed above the main diagonal, and if $k < 0$, they are placed below the main diagonal.

EXAMPLE 2.10:

For example, enter the following:

```
V = [ 2, 4, 6, 8 ];
diag(V, 0);
```

The following results will be displayed:

```
ans =
        2     0     0     0
        0     4     0     0
        0     0     6     0
        0     0     0     8
```

PRACTICE!

This practice problem will help you understand the special values and special matrices described in this section. Enter each of the following commands listed below, and observe how the matrix is changed:

```
C = diag(ones(4,1),-1)
C = C + diag(ones(4,1),1)
C = C + diag(ones(2,1),3)
C = C + diag(ones(2,1),-3)

ans =
        0     1     0     1     0
        1     0     1     0     1
        0     1     0     1     0
        1     0     1     0     1
        0     1     0     1     0
```

2.5 ADDITIONAL PLOTTING CAPABILITIES

The most common plot used by engineers and scientists is the xy plot. The data that we plot are usually read from a data file or computed in our programs and stored in vectors that we will call x and y. We generally assume that the x values represent the **independent variable** and that the y values represent the **dependent variable**. The y values can be computed as a function of x, or the x and y values might be measured in an experiment. We now present some additional ways of displaying this information.

2.5.1   Plot Commands

Most plots that we generate assume that the x- and y-axes are divided into equally spaced intervals; these plots are called **linear plots**. Occasionally, we may like to use a **logarithmic scale** on one or both of the axes. A logarithmic scale (base 10) is convenient when a variable ranges over many orders of magnitude, because the wide range of values can be graphed without compressing the smaller values.

The MATLAB commands for generating linear and logarithmic plots of the vectors **x** and **y** are the following:

plot(x,y)   Generates a linear plot of the values of **x** and **y**.

semilogx(x,y)   Generates a plot of the values of **x** and **y** using a logarithmic scale for **x** and a linear scale for **y**.

semilogy(x,y)   Generates a plot of the values of **x** and **y** using a linear scale for **x** and a logarithmic scale for **y**.

loglog(x,y)   Generates a plot of the values of **x** and **y** using a logarithmic scale for both **x** and **y**.

It is important to recognize that the logarithm of a negative value and of zero does not exist. Therefore, if the data to be plotted in a semilog plot or log–log plot contain negative values or zeros, a warning message will be printed by MATLAB informing you that these data points have been omitted from the data plotted.

Each of these commands can also be executed with one argument, as in **plot(y)**. In these cases, the plots are generated with the values of the indices of the vector **y** used as the **x** values.

2.5.2   Multiple Plots

One method for plotting multiple curves on the same graph is to use the **hold** command. Another method is to use multiple arguments in a plot command, as in

```
plot(X, Y, W, Z)
```

where the variables **X**, **Y**, **W**, and **Z** are vectors. When this command is executed, the curve corresponding to **X** versus **Y** will be plotted, and then the curve corresponding to **W** versus **Z** will be plotted on the same graph. The advantage of this technique is that MATLAB will automatically select different line types and colors, so that you can distinguish between the two plots.

If the plot function is called with a single matrix argument, MATLAB draws a separate line for each column of the matrix. The x-axis is labeled with the row index vector, $1:k$, where k is the number of rows in the matrix. If **plot** is called with two arguments, one a vector and the other a matrix, MATLAB successively plots a line for each row (or column) in the matrix.

EXAMPLE 2.11:   For example, enter the following statements to create data for a plot:

```
X  = 0.0:pi/100:2*pi;
Y1 = cos(X)*2;
Y2 = cos(X)*3;
Y3 = cos(X)*4;
Y4 = cos(X)*5;
Z  = [Y1; Y2; Y3; Y4];
```

Enter the following **plot** statement, and observe the results:

```
plot(X, Y1, X, Y2, X, Y3, X, Y4)
```

The following **plot** statement using the matrix argument **Z** will produce the same results:

```
plot(X, Z)
```

PRACTICE!

The **peaks** function is a function of two variables that produces sample data which are useful for demonstrating certain graphing functions. The data are created by scaling and translating Gaussian distributions. Calling **peaks** with a single argument n will create an $n \times n$ matrix. Use **peaks** to demonstrate the power of using a matrix argument to the **plot** function. Type the following statement to view the resulting plot:

```
plot(peaks(100))
```

2.5.3   Line and Mark Style

The command **plot(x,y)** generates a line plot that connects the points represented by the vectors **x** and **y** with line segments. You can also select other line types: dashed, dotted, and dash–dot. You can also select a point plot instead of a line plot. With a point plot, the points represented by the vectors will be marked with a point instead of being connected by line segments. You can also select characters other than a point to indicate the points. The other choices include plus signs, stars, circles, and x-marks. Table 2.6 contains these different options for lines and marks.

The following command illustrates the use of line and mark styles:

```
plot(x,y,x,y,'o')
```

It generates a solid line plot of the points represented by the vectors **x** and **y** and then plots the points themselves with circles. This type of plot is shown in Figure 2.4. The command

```
plot(x,y,'-o')
```

has a similar result and is easier to type.

TABLE 2.6   Line and Mark Options.

LINE TYPE	INDICATOR	POINT TYPE	INDICATOR
solid	-	point	.
dotted	:	circle	o
dash–dot	-.	x-mark	x
dashed	--	plus	+
		star	°
		square	s
		diamond	d
		triangle (down)	v
		triangle (up)	^
		triangle (left)	<
		triangle (right)	>
		pentagram	p
		hexagram	h

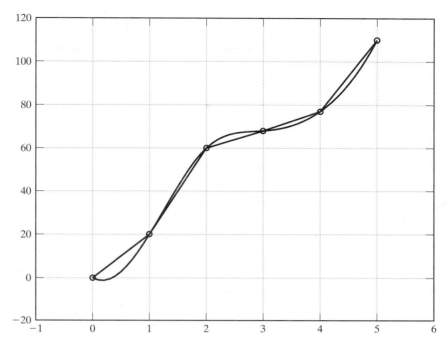

Figure 2.4.   Line and mark styles.

2.5.4   Axes Scaling

MATLAB automatically scales the axes to fit the data values. However, you can override this scaling with the **axis** command. There are several forms of the **axis** command:

> **axis**   Freezes the current axis scaling for subsequent plots. A second execution of the command returns the system to automatic scaling.
>
> **axis(v)**   Specifies that the axis being used is a four-element vector **v** that contains the scaling values **[xmin,xmax,ymin,ymax]**.

These commands are especially useful when you want to compare curves from different plots, because it can be difficult to visually compare curves plotted on different axes. The **plot** command precedes the corresponding **axis** command.

2.5.5   Subplots

The **subplot** command allows you to split the graph window into **subwindows**. Two subwindows can be arranged as either top and bottom or left and right. A four-window split has two subwindows on the top and two subwindows on the bottom. The arguments to the **subplot** command are three integers m, n, and p. The digits m and n specify that the graph window is to be split into an m-by-n grid of smaller windows, and the digit p specifies the pth window for the current plot. The windows are numbered from left to right, top to bottom. Therefore, the following commands specify that the graph window is to be split into a top plot and a bottom plot, and the current plot is to be placed in the top subwindow:

```
subplot(2,1,1), plot(x,y)
```

Figure 2.5 contains four plots that illustrate the **subplot** command, along with the linear and logarithmic plot commands. This figure was generated using the following statements:

```
%  Generate plots of a polynomial.
%
x = 0:0.5:50;
y = 5*x.^2;
subplot(2,2,1),plot(x,y),
   title('Polynomial - linear/linear'),
   ylabel('y'),grid,
subplot(2,2,2),semilogx(x,y),
   title('Polynomial - log/linear'),
   ylabel('y'),grid,
subplot(2,2,3),semilogy(x,y),
   title('Polynomial - linear/log'),
   xlabel('x'),ylabel('y'),grid,
subplot(2,2,4),loglog(x,y),
   title('Polynomial - log/log'),
   xlabel('x'),ylabel('y'),grid
```

Another example of the **subplot** command is included in the next section.

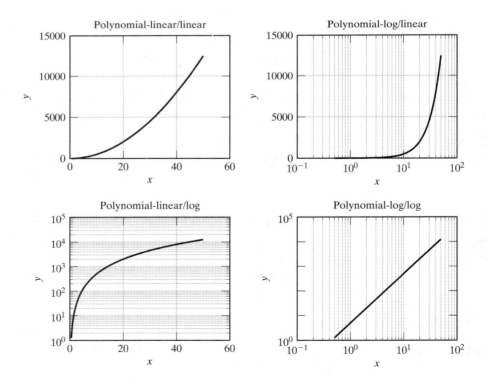

Figure 2.5.   Linear and logarithmic plots.

2.6   PROBLEM SOLVING APPLIED: VELOCITY COMPUTATION

In this section, we perform computations in an application related to the vehicle performance grand challenge. An advanced turboprop engine called an **unducted fan** (UDF) is one of the promising new propulsion technologies being developed for future transport aircraft. Turboprop engines, which have been in use for decades, combine the power and reliability of jet engines with the efficiency of propellers. They are a significant improvement over earlier piston-powered propeller engines. Their application has been limited to smaller commuter-type aircraft, however, because they are not as fast or powerful as the fanjet engines used on larger airliners. The UDF engine employs significant advancements in propeller technology, which narrow the performance gap between turboprops and fanjets. New materials, blade shapes, and higher rotation speeds enable UDF-powered aircraft to fly almost as fast as fanjets, and with greater fuel efficiency. The UDF is also significantly quieter than the conventional turboprop.

During a test flight of a UDF-powered aircraft, the test pilot has set the engine power level at 40,000 Newtons, which causes the 20,000-kg aircraft to attain a cruise speed of 180 meters/second. The engine throttles are then set to a power level of 60,000 Newtons, and the aircraft begins to accelerate. As the speed of the plane increases, the aerodynamic drag increases in proportion to the square of the airspeed. Eventually, the aircraft reaches a new cruise speed where the thrust from the UDF engines is just offset by the drag. The equations used to estimate the velocity and acceleration of the aircraft from the time that the throttle is reset until the time the plane reaches its new cruise speed (at approximately 120 s) are the following:

$$\text{velocity} \quad = 0.00001 \ \text{time}^3 - 0.00488 \ \text{time}^2$$
$$+ 0.75795 \ \text{time} + 181.3566$$
$$\text{acceleration} \quad = 3 - 0.000062 \ \text{velocity}^2$$

Write a MATLAB program that asks the user to enter a beginning time and an ending time (both in seconds) that define an interval of time over which we want to plot the velocity and acceleration of the aircraft. Assume that a time of zero represents the point at which the power level was increased. The ending time should be less than or equal to 120 seconds.

2.6.1   Problem Statement

Compute the new velocity and acceleration of the aircraft after a change in power level.

2.6.2   Input/Output Description

The input to the program is the starting and ending times, and the output of the program is a plot of the velocity and acceleration values over this window of time.

2.6.3   Hand Example

Because the program is generating a plot for a specified window of time, we will assume that the window is from 0 to 5 seconds. We then compute a few values with a calculator that can be compared with the values from the plot generated by the program:

TIME (s)	VELOCITY (m/s)	ACCELERATION (m/s^2)
0.0	181.3566	0.9608
3.0	183.5868	0.9103
5.0	185.0256	0.8775

2.6.4 Algorithm Development

The generation of the plot of the velocity and acceleration values requires the following steps:

1. Read time-interval limits.
2. Compute corresponding velocity and acceleration values.
3. Plot new velocity and acceleration values.

Because the time interval depends on the input values, it may be a very small interval or a very large interval. Therefore, instead of computing velocity and acceleration values at specified points, such as every 0.1 seconds, we will compute 100 points over the specified interval.

```
%  These commands generate and plot velocity and
%  acceleration values in a user-specified interval.
%
start_time = input('Enter start time (in seconds): ');
end_time = input('Enter ending time (max of 120 seconds): ');
%
time = linspace(start_time,end_time,100);
velocity = 0.00001*time.^3 - 0.00488*time.^2 ...
           + 0.75795*time + 181.3566;
acceleration = 3 - 0.000062*velocity.^2;
%
subplot(2,1,1),plot(time,velocity),title('Velocity'),
   ylabel('meters/second'),grid,
subplot(2,1,2),plot(time,acceleration),title('Acceleration'),
   xlabel('Time, s'),ylabel('meters/second^2'),grid;
```

2.6.5 Testing

We first test the program using the data from the hand example. This test generates the following interaction:

```
Enter start time (in seconds): 0
Enter ending time (max of 120 seconds): 5
```

The plot generated by the program is shown in Figure 2.6. Because the values computed match those from the hand example, we can test the program with other time values. If the values had not matched those from the hand example, we would need to determine if the error was in the hand example or in the program. The plot generated for the time interval from 0 to 120 seconds is shown in Figure 2.7. Note that the acceleration approaches zero as the velocity approaches its new cruise speed.

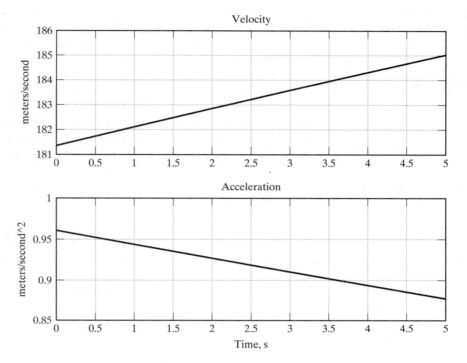

Figure 2.6.   Plot of hand example.

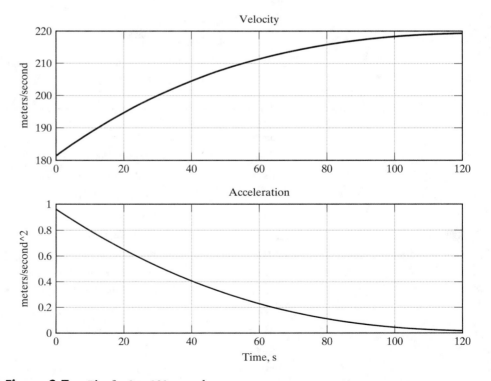

Figure 2.7.   Plot for 0 to 120 seconds.

SUMMARY

In this chapter, we have introduced you to the MATLAB environment. The primary data structure in MATLAB is a matrix, which can be a single point (a scalar), a list of values (a vector), or a rectangular grid of values with rows and columns. Values can be entered into a matrix by explicitly listing the values. Values can also be loaded into a matrix from MAT-files or ASCII files. In addition, values can be entered into a matrix using the colon operator, which allows us to specify a starting value, an increment, and an ending value to generate the sequence of values. We explored the various mathematical operations that are performed in an element-by-element manner. After examining a number of examples, we demonstrated how to generate simple xy plots of data values. The chapter closed with an application that plotted data from an experimental aircraft engine.

MATLAB SUMMARY

This MATLAB summary lists all the special characters, commands, and functions that were defined in this chapter.

Special Characters

[ ]	forms matrices
( )	forms subscripts
,	separates subscripts or matrix elements
;	separates commands or matrix rows
%	indicates comments
:	generates matrices
+	scalar and array addition
–	scalar and array subtraction
*	scalar multiplication
.*	array multiplication
/	scalar division
./	array division
^	scalar exponentiation
.^	array exponentiation

Commands and Functions

ans	stores expression values
axis	controls axis scaling
^c	generates a local abort
clc	clears command screen
clear	clears workspace
clf	clears figure
clock	represents the current time
date	represents the current date
demo	runs demonstrations
diag	creates a diagonal matrix or extracts a diagonal from a matrix
disp	displays matrix or text
eps	represents floating-point precision
eye	generates identity matrix

Commands and Functions

exit	terminates MATLAB
format +	sets format to plus and minus signs only
format compact	sets format to compact form
format long	sets format to long decimal
format long e	sets format to long exponential
format loose	sets format to noncompact form
format short	sets format to short decimal
format short e	sets format to short exponential
fprintf	prints formatted information
grid	inserts grid in a plot
help	invokes help facility
hold on	retains the current plot and its properties
i	represents the value $\sqrt{-1}$
Inf	represents the value ∞
input	accepts input from the keyboard
j	represents the value $\sqrt{-1}$
linspace	linearly spaced vector
load	loads matrices from a file
loglog	generates a log–log plot
NaN	represents the value Not-a-Number
ones	generates matrix of ones
pause	temporarily halts a program
peaks	generates sample data
pi	represents the value π
plot	generates a linear xy plot
print	prints the graphics window
quit	terminates MATLAB
randn	generates a Gaussian random number
save	saves variables in a file
semilogx	generates a log–linear plot
semilogy	generates a linear–log plot
size	determines row and column dimensions
subplot	splits graphics window into subwindows
title	adds a title to a plot
who	lists variables in memory
whos	lists variables and their sizes
xlabel	adds x-axis label to a plot
ylabel	adds y-axis label to a plot
zeros	generates matrix of zeros

KEY TERMS

arguments
command window
dependent variable
edit window
exponent overflow
exponent underflow

graphics window
independent variable
linear plots
logarithmic scale
main diagonal
matrix

prompt
scientific notation
submatrix
subscripts
transpose

Problems

1. Using the online help, determine and state the limitations of the **eye** command.

2. Which of the following are legitimate variable names in MATLAB?

   3vars
   global
   help
   My_var
   sin
   X+Y
   _input
   input

   Test your answers by trying to assign a value to each name—for example, **foo =3**.

3. Create the following matrix **A**:

$$A = \begin{bmatrix} 3.4 & 2.1 & 0.5 & 6.5 & 4.2 \\ 4.2 & 7.7 & 3.4 & 4.5 & 3.9 \\ 8.9 & 8.3 & 1.5 & 3.4 & 3.9 \end{bmatrix}$$

   Show the commands that return the following (use the colon operator):

   a. the first, second, and third columns
   b. the first, third and fifth columns
   c. the first-, third-, and fifth-column entries of the first and third rows
   d. the main diagonal (3.4, 7.7, and 1.5)
   e. the diagonal that contains the values (0.5, 4.5, 3.9)

4. Verify that **eps** is the smallest value by which two numbers can differ:

   ```
   a = 1;
   a == a %displays 1 if true, 0 if false
   a == a + eps %displays 1 if true, 0 if false
   a == a + eps/2 %displays 1 if true, 0 if false
   ```

5. Generate a vector of values from zero to 2π in increments of $\pi/100$. How can you determine the size of the vector that is generated?

6. Print the specified tables using the transpose operator as needed. Include a table heading and column headings. Choose an appropriate number of decimal places to print.

- Generate a table of conversions from degrees to radians. The first line should contain the values for 0°, the second line should contain the values for 10°, and so on. The last line should contain the values for 360°.
- Generate a table of conversions from centimeters to inches. Start the centimeters column at 0 and increment by 2 cm. The last line should contain the value 50 cm. (Recall that 1 in = 2.54 cm)
- Generate a table of conversions from mi/hr to ft/s. Start the mi/hr column at 0, and increment by 5 mi/hr. The last line should contain the value 65 mi/hr. (Recall that 1 mi = 5280 ft).

7. Use your favorite Internet search engine and World Wide Web browser to identify recent currency conversions for Finnish markkaa, Japanese yen, German Deutsche marks (DM), and U.S. dollars. Use the conversion values to complete the following tasks.

- Generate a table of conversions from markkaa to dollars. Start the markkaa column at 5 markkaa and increment by 5 markka. Print 25 lines in the table.
- Generate a table of conversions from DM to markkaa. Start the DM column at 1 DM and increment by 2 DM. Print 30 lines in the table.
- Generate a table of conversions from yen to DM. Start the yen column at 100 ¥, and print 25 lines, with the final line containing the value 10,000 ¥.

8. This set of problems requires you to generate temperature conversion tables. Use the following equations, which describe the relationships between temperatures in degrees Fahrenheit (T_F), degrees Celsius (T_C), degrees Kelvin (T_K), and degrees Rankine (T_R), respectively:

$$T_F = T_R - 459.67°\text{R}$$

$$T_F = \frac{9}{5}T_C + 32°\text{F}$$

$$T_R = \frac{9}{5}T_K$$

- Generate a table with the conversions from Fahrenheit to Kelvin for values from 0°F to 200°F. Allow the user to enter the increments in degrees F between lines.
- Generate a table with the conversions from Celsius to Rankine. Allow the user to enter the starting temperature and increment between lines. Print 25 lines in the table.
- Generate a table with conversions from Celsius to Fahrenheit. Allow the user to enter the starting temperature, the increment between lines, and the number of lines for the table.

9. The general equation for the distance that a free-falling body has traveled (neglecting air friction) is as follows:

$$d = \frac{1}{2}gt^2$$

Assume that $g = 9.8$ m/s². Plot the distance versus time for $t = 0$ to 100 seconds. Create two subplots, using linear coordinates and log–log coordinates, respectively. Label the axes, and give titles to the plots.

10. A function can be plotted in polar coordinates using the MATLAB statement

    ```
    polar(theta, sigma)
    ```

    where **theta** is the angle in radians, and **sigma** is the radius. Use the polar function to plot

    ```
    y=sin(x)
    ```

    at increments of $\pi/100$ between 0 and 2π.

11. The MATLAB function **randn(n,1)** will create an $n \times 1$ matrix that contains random entries, chosen from a normal distribution with mean zero and variance one. Use this function to generate four subplots of

    ```
    y=sin(x)
    ```

    that show the progressive addition of noise to y.

12. The forward kinematics map for a robot arm maps the joint angles to the position of the hand in the workspace. A simple planar robot with two degrees of freedom might have a kinematic map that would be represented in code as

    ```
    elbow_x = cos(theta1)
    elbow_y = sin(theta1)
    hand_x = elbow_x + cos(theta1 + theta2)
    hand_y = elbow_y + sin(theta2 + theta2)
    ```

    Write a program that will allow the user to select two joint angles with the mouse by using the **ginput** command to select a point from a figure. Then use this input data point to draw the arm in a second figure by plotting the base point of the arm at the origin and then the elbow and hand. Remember to use radians for the computation. Get help on the commands **figure** and **ginput** to find out more about them for this problem.

3

MATLAB Functions

GRAND CHALLENGE: SPEECH RECOGNITION

The modern jet cockpit has literally hundreds of switches and gauges. Several research programs have been looking at the feasibility of using a speech recognition system in the cockpit to serve as a pilot's assistant. The system would respond to verbal requests from the pilot for information such as fuel status or altitude. The pilot would use words from a small vocabulary that the computer had been trained to understand. In addition to understanding a selected set of words, the system would also have to be trained to understand the speech of the particular pilot who would be using the system. This training information could be stored on a diskette and inserted into the computer at the beginning of a flight so that the system could recognize the current pilot. The computer system would also use speech synthesis to respond to the pilot's request for information.

SECTIONS

OBJECTIVES

After reading this chapter, you should be able to

- apply a variety of mathematical and trigonometric functions to matrices
- compute descriptive statistics and plot histograms
- generate uniform and Gaussian random sequences
- apply the principles in this chapter to a flight simulation problem
- understand the use of logical operators, logical functions, and control structures in MATLAB, and
- write simple MATLAB programs and user-defined functions.

3.1   MATH FUNCTIONS AND TRIGONOMETRIC FUNCTIONS

Arithmetic expressions often require computations other than addition, subtraction, multiplication, division, and exponentiation. For example, many expressions require the use of logarithms, exponentials, and trigonometric functions. MATLAB allows us to use function references to perform these types of computations instead of requiring us to compute them using the basic arithmetic operations. For example, if we want to compute the sine of an angle and store the result in **b**, we can use the following command:

```
b = sin(angle);
```

The **sin** function assumes that the argument is in radians. If the argument contains a value in degrees, we can convert the degrees to radians within the function reference:

```
b = sin(angle*pi/180);
```

We could also have done the conversion in a separate statement:

```
angle_radians = angle*pi/180;
b = sin(angle_radians);
```

These statements are valid if **angle** is a scalar or if **angle** is a matrix. If **angle** is a matrix, then the function will be applied element by element to the values in the matrix.

Now that you have seen an example using a function, we discuss the rules regarding them. A **function** is a reference that represents a matrix. The arguments, or parameters, of the function are contained in parentheses following the name of the function. A function may contain no arguments, one argument, or many arguments, depending on its definition. For example, **pi** is a function that has no argument. When we use the function reference **pi**, the value for π automatically replaces the function reference. If a function contains more than one argument, it is very important to give the arguments in the correct order. Some functions also require that the arguments be in specific units. For example, the trigonometric functions assume that the arguments are in radians. In MATLAB, some functions use the number of arguments to determine the output of the function. Also, note that the names of built-in functions are in lowercase.

A function reference cannot be used on the left side of an equals sign, because it represents a value, not a variable. Functions can appear on the right side of an equals sign and in expressions. A function reference can also be part of the argument of another function reference. For example, the following statement computes the logarithm of the absolute value of x:

```
log_x = log(abs(x));
```

When one function is used to compute the argument of another function, be sure to enclose the argument of each function in its own set of parentheses. This **nesting of functions** is also called **composition of functions**.

We now discuss several categories of functions that are commonly used in engineering computations. Other functions will be presented throughout the remaining chapters as we discuss relevant subjects. Tables of common functions are included on the final two pages of this book.

3.1.1   Elementary Math Functions

The elementary math functions include functions to perform a number of common computations, such as computing the absolute value and the square root of a number. In addition, we also include a group of functions used to perform rounding. We now list and briefly describe these functions:

abs(x)   Computes the absolute value of **x**.

sqrt(x)   Computes the square root of **x**.

round(x)   Rounds **x** to the nearest integer.

fix(x)   Rounds (or truncates) **x** to the nearest integer toward zero.

floor(x)   Rounds **x** to the nearest integer toward $-\infty$.

ceil(x)   Rounds **x** to the nearest integer toward ∞.

sign(x)   Returns a value of -1 if **x** is less than zero, a value of zero if **x** equals zero, and a value of 1 otherwise.

rem(x,y)   Computes the remainder of $\frac{x}{y}$. For example, **rem(25,4)** is 1, and **rem(100,21)** is 16.

exp(x)   Computes the value of e^x, where e is the base for natural logarithms, or approximately 2.718282.

log(x)   Computes ln **x**, the natural logarithm of **x** to the base e.

log10(x)   Computes $\log_{10}$ **x**, the common logarithm of **x** to the base 10.

PRACTICE!

Evaluate the following expressions, and then check your answer by entering the expressions in MATLAB:

a.   **round(-2.6)**

b.   **fix(-2.6)**

c.   **floor(-2.6)**

d.   **ceil(-2.6)**

e.   **sign(-2.6)**

f.   **rem(15,2)**

g.   **floor(ceil(10.8))**

h.   **log10(100) + log10(0.001)**

i.   **abs(-5:5)**

j.   **round([0:0.3:2,1:0.75:4])**

3.1.2   Trigonometric Functions

The trigonometric functions assume that angles are represented in radians. To convert radians to degrees or degrees to radians, use the following conversions, which use the fact that $180° = \pi$ radians:

```
angle_degrees = angle_radians*(180/pi);
angle_radians = angle_degrees*(pi/180);
```

We now list and briefly describe the trigonometric functions:

sin(x)   Computes the sine of **x**, where **x** is in radians.

cos(x)   Computes the cosine of **x**, where **x** is in radians.

tan(x)   Computes the tangent of **x**, where **x** is in radians.

asin(x)   Computes the arcsine, or inverse sine, of **x**, where **x** must be between −1 and 1. The function returns an angle in radians between $\dfrac{-\pi}{2}$ and $\dfrac{\pi}{2}$.

acos(x)   Computes the arccosine, or inverse cosine, of **x**, where **x** must be between −1 and 1. Returns an angle in radians between 0 and π.

atan(x)   Computes the arctangent, or inverse tangent, of **x**. Returns an angle in radians between $\dfrac{-\pi}{2}$ and $\dfrac{\pi}{2}$.

atan2(y,x)   Computes the arctangent, or inverse tangent, of the value $\dfrac{x}{y}$. Returns an angle in radians, which will be between $-\pi$ and π, depending on the signs of **x** and **y**.

The other trigonometric functions can be computed using the following equations:

$$\sec x = \frac{1}{\cos x} \qquad \csc x = \frac{1}{\sin x} \qquad \cot x = \frac{1}{\tan x}$$

PRACTICE!

Give MATLAB commands for computing the following values:

1. Uniformly accelerated motion:

$$\text{motion} = \sqrt{vi^2 + 2 \cdot a \cdot x}$$

2. Electrical oscillation frequency:

$$\text{frequency} = \frac{1}{\sqrt{\dfrac{2\pi c}{L}}}$$

3. Range for a projectile:

$$\text{range} = 2vi^2 \cdot \frac{\sin(b) \cdot \cos(b)}{g}$$

4. Length contraction:

$$\text{length} = k\sqrt{1 - \left(\frac{y}{c}\right)^2}$$

5. Volume of a fillet ring:

$$\text{volume} = 2\pi x^2 \left(\left(1 - \frac{\pi}{4}\right) \cdot y - \left(0.8333 - \frac{\pi}{4}\right) \cdot x \right)$$

6. Distance of the center of gravity from a reference plane in a sector of a hollow cylinder:

$$\text{center} = \frac{38.1972 \cdot (r^3 - s^3)\sin a}{(r^2 - s^2) \cdot a}$$

3.2   DATA ANALYSIS FUNCTIONS

Analyzing data is an important part of evaluating test results. MATLAB contains a number of functions that make it easier to evaluate and analyze data. We first present a number of simple analysis functions, and then we present functions that compute more complicated measures or metrics related to a data set.

3.2.1   Simple Analysis

The following groups of functions are frequently used in evaluating a set of test data:

Maximum and Minimum   This set of functions can be used to determine maximums and minimums and their locations.

> **max(x)**   Returns the largest value in a vector **x**. Returns a row vector containing the maximum element from each column of a matrix **x**.
>
> **max(x,y)**   Returns a matrix the same size as **x** and **y**. Each element in the matrix contains the maximum value from the corresponding positions in **x** and **y**.
>
> **min(x)**   Returns the smallest value in a vector **x**. Returns a row vector containing the minimum element from each column of a matrix **x**.
>
> **min(x,y)**   Returns a matrix the same size as **x** and **y**. Each element in the matrix contains the minimum value from the corresponding positions in **x** and **y**.

Mean and Median   The **mean** of a group of values is the average of the values. The Greek symbol μ (mu) is used to represent the value of the mean, as shown in the following equation, which uses summation notation to define the mean:

$$\mu = \frac{\sum_{k=1}^{N} x_k}{N}$$

Here $\sum_{k=1}^{N} x_k = x_1 + x_2 + \cdots + x_N$.

The **median** is the value in the middle of the group, assuming that the values are sorted. If there is an odd number of values, the median is the value in the middle position. If there is an even number of values, then the median is the mean of the two middle values.

The functions for computing the mean and median are the following:

> **mean(x)**   Computes the mean value (or average value) of the elements of a vector **x**. Computes a row vector that contains the mean value of each column of a matrix **x**.
>
> **median(x)**   Computes the median value of the elements in a vector **x**. Computes a row vector that contains the median value of each column of a matrix **x**.

Sums and Products   MATLAB contains functions for computing the sums and products of columns in a matrix and functions for computing the cumulative sums and products of the elements in a matrix.

sum(x)   Computes the sum of the elements in a vector **x**. Computes a row vector that contains the sum of each column of a matrix **x**.

prod(x)   Computes the product of the elements in a vector **x**. Computes a row vector that contains the product of each column of a matrix **x**.

cumsum(x)   Computes a vector of the same size containing cumulative sums of values from a vector **x**. Computes a matrix the same size as **x** containing cumulative sums of values from the columns of **x**.

cumprod(x)   Computes a vector of the same size containing cumulative products of values from a vector **x**. Computes a matrix the same size as **x** containing cumulative products of values from the columns of **x**.

Sorting Values   MATLAB contains a function for sorting values into ascending order.

sort(x)   Sorts the values of a vector **x** into ascending order. Sorts each column of a matrix **x** into ascending order.

PRACTICE!

Determine the matrices represented by each function reference. Then use MATLAB to check your answers. Assume that **w**, **x**, and **y** are the following matrices:

$$\mathbf{w} = \begin{bmatrix} 0 & 3 & -2 & 7 \end{bmatrix} \quad \mathbf{x} = \begin{bmatrix} 3 & -1 & 5 & 7 \end{bmatrix}$$

$$\mathbf{y} = \begin{bmatrix} 1 & 3 & 7 \\ 2 & 8 & 4 \\ 6 & -1 & -2 \end{bmatrix}$$

a.   **max(w)**
b.   **min(y)**
c.   **min(w,x)**
d.   **mean(y)**
e.   **median(w)**
f.   **cumprod(y)**
g.   **sort(2*w+x)**
h.   **sort(y)**

3.2.2   Variance and Standard Deviation

Two of the most important statistical measurements of a set of data are the variance and standard deviation. Before we give their mathematical definitions, it is useful to develop an intuitive understanding of these values. Consider the values of vectors **data_1** and **data_2**, plotted in Figure 3.1. If we attempt to draw a line through the middle of the values in the plots, this line would be at approximately 3.0 in both plots. Thus, we would assume that both vectors have approximately the same mean value of 3.0. However, the data in the two vectors clearly have some distinguishing characteristics. The data values in **data_2** vary more, or deviate more, from the mean. Thus, measures of variance and deviation for the values in **data_2** will be greater than measures of variance and deviation for the values in **data_1**. Hence, an intuitive understanding of variance (or deviation) relates to the variance of the values from the mean. The larger the variance, the further the values fluctuate from the mean value.

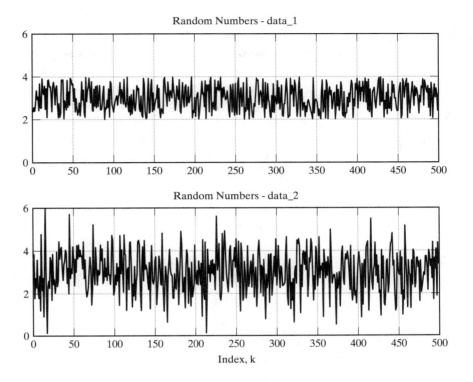

Figure 3.1. Random sequences.

Mathematically, the **variance** σ^2 for a set of data values (which we will assume are stored in a vector x) can be computed using the following equation:

$$\sigma^2 = \frac{\sum\limits_{k=1}^{N}(x_k - \mu)^2}{N-1}$$

This equation is a bit intimidating at first, but if you look at it closely, it seems much simpler. The term $x_k - \mu$ is the difference, or deviation, of x_k from the mean. This value is squared so that we will always have a positive value. We then add together the squared deviations for all the data points. This sum is then divided by $N - 1$, which approximates an average. (The equation for the variance sometimes uses a denominator of N, but the form here has statistical properties that make it generally more desirable.) Thus, the variance is the average squared deviation of the data from the mean.

The **standard deviation** is defined as the square root of the variance, or

$$\sigma = \sqrt{\sigma^2}$$

where σ is the Greek symbol sigma. MATLAB includes a function to compute the standard deviation. To compute the variance, simply square the standard deviation.

std(x) Computes the standard deviation of the values in a vector **x**.
Computes a row vector containing the standard deviation of each column of a matrix **x**.

3.2.3 Histograms

The **histogram** is a special type of graph that is particularly relevant to the statistical measurements discussed in this section. A histogram is a plot showing the distribution of a set of values. In MATLAB, the histogram computes the number of values falling in 10 bins that are equally spaced between the minimum and maximum values from the set of values. For example, if we plot the histograms of the data values in vectors **data_1** and **data_2** (see Figure 3.1), we obtain the histograms in Figure 3.2. Note that the information obtained from a histogram is different than the information obtained from the mean and variance. The histogram shows us not only the range of values, but also how they are distributed. For example, the values in **data_1** tend to be equally distributed across the range of values. (In Section 3.3, we will see that these types of values are called *uniformly distributed values.*) The values in **data_2** are not equally distributed across the range of values. In fact, most of the values are centered around the mean. (In Section 3.3, we will see that this type of distribution is called a *Gaussian*, or *normal distribution*.)

The MATLAB command to generate and plot a histogram is

hist(x)

where **x** is a vector containing the data to be used in the histogram. The **hist** command also allows us to select the number of bins. Therefore, if we want to increase the resolution of the histogram so that 25 bins are used, instead of 10, we use the following command:

hist(x,25)

The corresponding plots with 25 bins, using the **data_1** and **data_2** vectors, are shown in Figure 3.3.

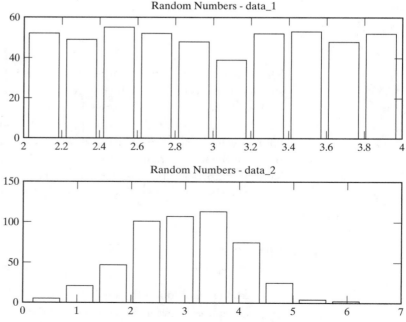

Figure 3.2. Histograms with 10 bins.

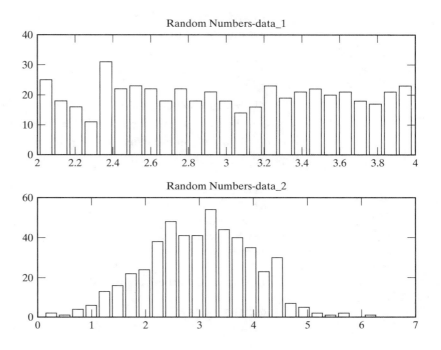

Figure 3.3. Histograms with 25 bins.

3.3 RANDOM NUMBERS

There are many engineering problems that require the use of random numbers in the development of a solution. In some cases, the random numbers are used to develop a **simulation** of a complex problem. The simulation can be tested over and over to analyze the results, and each test represents a repetition of the experiment. We also use random numbers to approximate noise sequences. For example, the static that we hear on a radio is a noise sequence. If we are testing a program that uses an input data file which represents a radio signal, we may want to generate noise and add it to a speech signal or a music signal in order to provide a more realistic signal.

3.3.1 Uniform Random Numbers

Random numbers are not defined by an equation; instead, they can be characterized by a distribution of values. For example, random numbers that are equally likely to be any value between an upper and lower limit are called **uniform random numbers**. The top histogram in Figure 3.3 shows the distribution of a set of uniform values between 2 and 4.

The **rand** function in MATLAB generates random numbers uniformly distributed over the interval [0,1]. A **seed** value is used to initiate a random sequence of values. This seed value is initially set to zero, but it can be changed with the **rand** function.

rand(n) Returns an **n** × **n** matrix. Each value in the matrix is a random number between 0 and 1.

rand(m,n) Returns an **m** × **n** matrix. Each value in the matrix is a random number between 0 and 1.

rand('seed',n) Sets the value of the seed to the value n. The value of n is initially set to 0.

rand('seed') Returns the current value of the random-number generator.

The **rand** function generates the same sequence of random values in each work session if the same seed value is used. The commands shown next generate and print two sets of 10 random numbers uniformly distributed between 0 and 1. The difference between the two sets is caused by the different seeds:

```
rand('seed',0)
set1 = rand(10,1);
rand('seed',123)
set2 = rand(10,1);
[set1 set2]
```

The values printed by these commands are

```
0.2190 0.0878
0.0470 0.6395
0.6789 0.0986
0.6793 0.6906
0.9347 0.3415
0.3835 0.2359
0.5194 0.2641
0.8310 0.6044
0.0346 0.4181
0.0535 0.1363
```

Random sequences with values that range between values other than 0 and 1 are often needed. To illustrate this point, suppose that we want to generate values between −5 and 5. We first generate a random number r (which is between 0 and 1) and then multiply it by 10, which is the difference between the upper and lower bounds (5 − (−5)). We then add the lower bound (−5), giving a resulting value that is equally likely to be any value between −5 and 5. Thus, if we want to convert a value r that is uniformly distributed between 0 and 1 to a value uniformly distributed between a lower bound a and an upper bound b, we use the following equation:

$$x = (b - a) \cdot r + a$$

The sequence **data_1**, plotted in Figure 3.1, was generated with this equation:

```
data_1 = rand(1,500)*2 + 2;
```

Thus, the sequence contains 500 values uniformly distributed between 2 and 4. The random-number seed was 246.

PRACTICE!

Give the MATLAB statements to generate 10 random numbers within the specified range. Check your answers by executing the statements and printing the values generated in the vectors.

  a.   Uniform random numbers between 0 and 10.0.
  b.   Uniform random numbers between −1 and +1.
  c.   Uniform random numbers between −20 and −10.
  d.   Uniform random numbers between 4.5 and 5.0.
  e.   Uniform random numbers between −π and π.

3.3.2 Gaussian Random Numbers

When we generate a random-number sequence with a uniform distribution, all values are equally likely to occur. However, we sometimes need to generate random numbers using distributions in which some values are more likely to be generated than others. For example, suppose that a random-number sequence represents outdoor temperature measurements taken over a period of time. We would find that the temperature measurements have some variation, but typically are not equally likely. For example, we might find that the values vary over only a few degrees, although larger changes could occasionally occur because of storms, cloud shadows, and day-to-night changes.

Random-number sequences that have some values which are more likely to occur than others can often be modeled with a **Gaussian random variable** (also called a **normal random variable**). An example of a set of values with a Gaussian distribution is shown in the second plot in Figure 3.3. The mean value of this random-number sequence corresponds to the x-coordinate of the peak of the distribution, which is approximately 3. From this histogram, you can see that most values are close to the mean. Although a uniform random variable has specific upper and lower bounds, a Gaussian random variable is not defined in terms of upper and lower bounds; it is defined in terms of the mean and variance of the values. For Gaussian random numbers, it can be shown that approximately 68% of the values will fall within one standard deviation of the mean, 95% will fall within two standard deviations of the mean, and 99% will fall within three standard deviations of the mean. These statistics are useful in working with Gaussian random numbers.

MATLAB will generate Gaussian values with a mean of 0 and a variance of 1.0 if we specify a normal distribution. The functions for generating Gaussian values are as follows:

randn(n)   Returns an **n** × **n** matrix. Each value in the matrix is a Gaussian (or normal) random number with a mean of 0 and a standard deviation of 1.

randn(m,n)   Returns an **m** × **n** matrix. Each value in the matrix is a Gaussian (or normal) random number with a mean of 0 and a standard deviation of 1.

To modify Gaussian values with a mean of 0 and a variance of 1 to another Gaussian distribution, multiply the values by the standard deviation of the desired distribution, and add the mean of the desired distribution. Thus, if r is a random number with a mean of 0 and a variance of 1.0, the following equation will generate a new random number with a standard deviation of a and a mean of b:

$$x = a \cdot r + b$$

The sequence **data_2**, plotted in Figure 3.1, was generated with the following equation:

```
data_2 = randn(1,500) + 3;
```

Thus, the sequence contains 500 Gaussian random variables with a standard deviation of 1 and a mean of 3. The random-number seed used was 95.

3.4 PROBLEM SOLVING APPLIED: FLIGHT SIMULATOR

Computer simulations are used to generate situations that model or emulate a real-world situation. Some computer simulations are written to play games such as checkers, poker, and chess. To play the game, you indicate your move, and the computer will select an appropriate response. Other animated games use computer graphics to develop an interaction as you use the keys or a mouse to play the game. In more sophisticated computer simulations, such as those in a flight simulator, the computer not only responds to the input from the user, but also generates values, such as temperatures, wind speeds, and the locations of other aircraft. The simulators also simulate emergencies that occur during the flight of an aircraft. If all of this information generated by the computer were always the same set of information, the value of the simulator would be greatly reduced. It is important that there be randomness to the generation of the data. Simulations that use random numbers to generate values that model events are called **Monte Carlo simulations**.

Write a program to generate a random-number sequence to simulate one hour of wind speed data that are updated every 10 seconds. Assume that the wind speed will be modeled as a uniform random number that varies between a lower limit and an upper limit. Plot the data, and save it in an ASCII file named **windspd.dat**.

3.4.1 Problem Statement

Generate one hour of simulated wind speed data using a lower limit and an upper limit.

3.4.2 Input/Output (I/O)

As shown in the accompanying I/O diagram, the inputs to the program are the lower limits and upper limits for the wind speed. The output is the plot and the data file containing the simulated wind speeds.

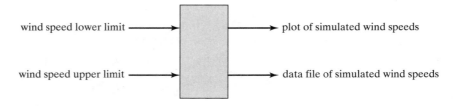

wind speed lower limit ⟶ ⟶ plot of simulated wind speeds

wind speed upper limit ⟶ ⟶ data file of simulated wind speeds

3.4.3 Hand Example

This simulation uses MATLAB's random-number generator to generate numbers between 0 and 1. We then modify these values to be between a specified lower limit and an upper limit. Thus, to generate a value between 10 and 15, we would multiply a random number (between 0 and 1) by 5 and add 10 to it. Hence, the value 0.1 would be converted to 10.5.

3.4.4   MATLAB Solution

```
%  These statements generate one hour of simulated
%  wind speeds based on inputs for the lower
%  and upper limits.
%
low_speed = input('Enter lower limit for wind speed: ');
high_speed = input('Enter upper limit for wind speed: ');
seed = input('Enter seed for random numbers: ');
%
%      Generate simulated wind speed.
%
t = 0:1/360:1;
rand('seed',seed)
speed = (high_speed - low_speed)*rand(1,361) + low_speed;
plot(t,speed),title('Simulated Wind Speed'),
xlabel('t,hours'),ylabel('wind, mi/hr'),grid,pause
data(:,1) = t';
data(:,2) = speed';
save windspd.dat data /ascii
```

3.4.5   Testing

Figure 3.4 contains plots of the wind speed, given the same limits (between 3 and 6 mi/hr) but with different seeds (seeds of 123 and 246).

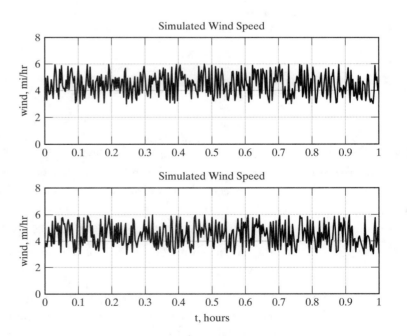

Figure 3.4.   Simulated wind speeds with different random-number seeds.

3.5 STATEMENT-LEVEL CONTROL STRUCTURES

The control flow statements in a program determine the order of computations. Two types of control flow statements are **selection statements** and loops. A selection statement provides the means of choosing from among two or more alternative paths in a program. The selection statement allows us to test a **logical condition** to determine which steps are to be performed next. The conditions that are evaluated often contain relational and logical operators, or they contain relational and logical functions. A loop control statement causes a group of statements to be executed zero, one, or more times. The number of times that the loop executes is dependent on a counter or a logical condition.

In this section, we first discuss the most common form of selection structure—the simple **if statement**—and then discuss relational and logical operators. We then discuss **nested if structures**, the MATLAB predefined logical functions, and finally the MATLAB loop structures: the **for and while statements**.

3.5.1 Simple *if* Statement

The simple if statement has the following form:

```
if expression
   statements
end
```

If the logical expression is true, the statements between the **if** statement and the **end** statement are executed. If the logical expression is false, program control jumps immediately to the statement following the **end** statement. It is good programming practice to indent the statements within an **if** structure, for readability.

EXAMPLE 3.1:   An example of an **if** statement is as follows:

```
if G < 50
  count = count + 1;
  disp (G);
end
```

Assume that **G** is a scalar. If **G** < 50, then **count** is incremented by 1, and **G** is displayed on the screen; otherwise, these two statements are skipped. If **G** is not a scalar, then **count** is incremented by 1, and **G** is displayed only if every element in **G** is less than 50.

Because logical expressions are generated from relational operators and logical operators, we now discuss these new operators.

3.5.2 Relational and Logical Operators

MATLAB has six relational operators for comparing two matrices of equal size, as shown in Table 3.1. Matrices or matrix expressions are used on both sides of a relational operator to yield another matrix of the same size. Each entry in the resulting matrix contains a 1 if the comparison is true when applied to the values in the corresponding position of the matrices; otherwise, the entry in the resulting matrix contains a 0. An expression that contains a relational operator is a logical expression, because the result is a matrix containing 0's and 1's, which can be interpreted as false values and true values, respectively; the resulting matrix is also called a **0–1 matrix**.

TABLE 3.1 Relational Operators.

RELATIONAL OPERATOR	INTERPRETATION
<	less than
<=	less than or equal to
>	greater than
>=	greater than or equal to
==	equal to
~=	not equal to

Consider the logical expression **a<b**. If **a** and **b** are scalars, the value of this expression is 1 (for true) if **a** is less than **b**; otherwise, the expression is 0 (for false). Let **a** and **b** be vectors with the following values:

$$a = \begin{bmatrix} 2 & 4 & 6 \end{bmatrix} \quad b = \begin{bmatrix} 3 & 5 & 1 \end{bmatrix}$$

Then, the value of **a<b** is the vector **[1 1 0]**, and the value of **a~=b** is **[1 1 1]**.

We can also combine two logical expressions, using the logical operators **not**, **and**, and **or**. These logical operators are represented by the symbols shown in Table 3.2. Logical operators allow us to compare 0–1 matrices, such as those computed by relational operators, as shown in the following logical expression:

```
a<b & b<c
```

TABLE 3.2 Logical Operators.

LOGICAL OPERATOR	SYMBOL
not	~
and	&
or		

This operation is valid only if the two resulting matrices (represented by **a<b** and **b<c**) are the same size. Then, an entry in the matrix represented by this logical expression is **1** if the values in the corresponding entries in **a**, **b**, and **c** are such that **a<b<c**; otherwise, the entry is **0**.

Actually, MATLAB supports *four* logical operators: **and** (**&**), **or** (**|**), **not** (**~**), and **xor**. If the **or** (**|**) operator is applied to two logical expressions, an entry in the resulting 0–1 matrix is **1** (true) if either of the expressions is true; it is **0** (false) only when both expressions are false. Any nonzero real value is considered true.

EXAMPLE 3.2:

For example, create the following matrices:

```
A = [0 2 -5];
B = [0 0  3];
```

Applying the logical or operator results in the following 0–1 matrix:

```
C = A | B
C =

    0    1    1
```

If the **and** (**&**) operator is applied to two logical expressions, an entry in the resulting 0–1 matrix is 1 (true) if both of the expressions are true; it is 0 (false) when either expression is false.

EXAMPLE 3.3:   Applying the logical and operator to **A** and **B** results in the following 0–1 matrix:

```
C = A & B
C =

    0    0    1
```

If the **xor** operator is applied to two logical expressions, an entry in the resulting 0–1 matrix is 1 (true) if either (but not both) of the expressions is true. It is 0 (false) when both the expressions are true or both are false.

EXAMPLE 3.4:   Applying the logical **xor** operator results in the following 0–1 matrix:

```
C = xor(A, B)
C =

    0    1    0
```

The binary operators (and, or, and xor) require that both expressions have the same dimensions, unless one of the expressions is a scalar. Table 3.3 lists the truth values for the four logical operators using two logical expressions.

TABLE 3.3 Truth Values of the Logical Operators.

A	B	~A	A|B	A & B	A xor B
0	0	1	0	0	0
0	1	1	1	0	1
1	0	0	1	0	1
1	1	0	1	1	0

The unary logical operator not can precede any logical expression. This operator changes the value of the expression to the opposite value; hence, if an element of **(A > B)** is true, then the same element of **~(A > B)** is false.

When a logical expression contains several logical operators, an order of precedence is followed. The order of precedence of the logical operators from highest to lowest is **not**, **and**, then **or**. Of course, parentheses can be used to change the hierarchy.

EXAMPLE 3.5:   For example, set the values of two variables **b** and **c** to be

```
b = 3;
c = 5;
```

Then the expression

```
~(b == c | b == 5.5)
```

evaluates to true. The expressions **b == c** and **b == 5.5** are evaluated first. Neither expression is true, so the expression **(b == c | b == 5.5)** is false. We then apply the ~ operator, which changes the value of the expression to true.

EXAMPLE 3.6:

As a second example, the expression

```
~b == c | b == 5.5
```

evaluates to false. In this case, the expression **~b == c** evaluates to false, and the expression **b == 5.5** evaluates to false. Thus, the value of the entire logical expression is false.

You might wonder how we could evaluate **~b** when the value in **b** is a number. In MATLAB, any values that are nonzero are considered to be true; values of zero are false. We have to be very careful using relational and logical operators to be sure that the steps being performed are the ones that we want to perform.

PRACTICE!

Determine if the expressions in Problems 1 through 8 are true or false. Then check your answers using MATLAB. Remember that to check your answers, all you need to do is enter the expression. Assume that the following variables have the indicated values:

```
    a = 5.5      b = 1.5      k = -3

    a.   a < 10.0
    b.   a+b >= 6.5
    c.   k ~= 0
    d.   b-k > a
    e.   ~(a == 3*b)
    f.   -k <= k+6
    g.   a<10 & a>5
    h.   abs(k)>3 | k<b-a
```

3.5.3   Nested **if** Statements

Here is an example of nested **if** statements that extends the previous example:

```
if g < 50
    count = count + 1;
    disp(g);
    if b > g
        b = 0;
    end
end
```

Again, first assume that **g** and **b** are scalars. Then, if **g < 50**, we increment **count** by 1 and display **g**. In addition, if **b > g**, then we also set **b** to zero. If **g** is not less than 50, then we skip immediately to the statement following the second **end** statement. If **g** is not a scalar, then the condition **g < 50** is true only if every element of **g** is less than 50. If neither **g** nor **b** is a scalar, then **b** is greater than **g** only if every corresponding pair of elements of **g** and **b** are values such that **b** is greater than **g**. If **g** or **b** is a scalar, then the other matrix is compared to the scalar element by element.

3.5.4 else and elseif Clauses

The **else** clause allows us to execute one set of statements if a logical expression is true and a different set of statements if the logical expression is false. To illustrate this feature, assume that we have a variable **interval**. If the value of **interval** is less than 1, we set the value of **x_increment** to **interval/10**; otherwise, we set the value of **x_increment** to **0.1**. The following statement performs these steps:

```
if interval < 1
    x_increment = interval/10;
else
    x_increment = 0.1;
end
```

When we nest several levels of **if-else** statements, it may be difficult to determine which logical expressions must be true (or false) to execute each set of statements. In these cases, the **elseif** clause is often used to clarify the program logic, as illustrated in the following statement:

```
if temperature > 100
    disp('Too hot - equipment malfunctioning.')
elseif temperature > 90
    disp('Normal operating range.')
elseif temperature > 50
    disp('Temperature below desired operating range.')
else
    disp('Too Cold - turn off equipment.')
end
```

In this example, temperatures above 90 and below or equal to 100 are in the normal operating range. Temperatures outside this range generate an appropriate message. Notice that a temperature of 101 does not trigger all of the responses.

PRACTICE!

In Problems 1 through 4, give MATLAB statements that perform the steps indicated. Assume that the variables are scalars.

a. If the difference between **volt_1** and **volt_2** is larger than 10.0, print the values of **volt_1** and **volt_2**.

b. If the natural logarithm of **x** is greater than or equal to 3, set **time** equal to 0 and increment **count**.

c. If **dist** is less than 50.0 and **time** is greater than 10.0, increment **time** by 2; otherwise, increment **time** by 2.5.

d. If **dist** is greater than or equal to 100.0, increment **time** by 2.0. If **dist** is between 50 and 100, increment **time** by 1. Otherwise, increment **time** by 0.5.

3.5.5 Logical Functions

MATLAB contains a set of logical functions that are very useful with **if** statements. We now discuss each of these functions.

any(x) Returns a scalar that is **1** (true) if any element in the vector **x** is not zero; otherwise, the scalar is **0** (false). Returns a row vector if **x** is a matrix. An element in this row vector contains a **1** (true) if any element of the corresponding column of **x** is nonzero, and a **0** (false) otherwise.

all(x) Returns a scalar that is **1** (true) if all elements in the vector **x** are not zero; otherwise, the scalar is **0** (false). Returns a row vector if **x** is a matrix. An element in this row vector contains a **1** (true) if all elements of the corresponding column of **x** are nonzero, and a **0** (false) otherwise.

find(x) Returns a vector containing the indices of the nonzero elements of a vector **x**. If **x** is a matrix, the indices are selected from **x(:)**, which is a long column vector formed from the columns of **x**.

isnan(x) Returns a matrix with ones when the elements of **x** are NaNs, and zeros when they are not.

finite(x) Returns a matrix with ones when the elements of **x** are finite, and zeros when they are infinite or NaN.

isempty(x) Returns 1 if **x** is an empty matrix, and zero otherwise.

Assume that **A** is a matrix with three rows and three columns of values. Consider the following statement:

```
if all(A)
     disp ('A contains no zeros')
end
```

The string **A contains no zeros** is printed only if all nine values in **A** are nonzero.

We now present another example that uses a logical function. Assume that we have a vector **d** containing a group of distance values that represent the distances of a cable car from the nearest tower. We want to generate a vector containing corresponding velocities of the cable car. If the cable car is within 30 feet of the tower, we use this equation to compute the velocity:

$$velocity = 0.425 + 0.00175d^2$$

If the cable car is further than 30 feet from the tower, we use the following equation:

$$velocity = 0.625 + 0.12d - 0.00025d^2$$

We can use the **find** function to find the distance values greater than 30 feet and to find the distance values less than or equal to 30 feet. Because the **find** function identifies the subscripts for each group of values, we can then compute the corresponding velocities with the following statements:

```
lower = find(d < 30);
velocity(lower) = 0.425 + 0.00175*d(lower).^2;
upper = find(d >= 30);
velocity(upper) = 0.625 + 0.12*d(upper)...
                  - 0.00025*d(upper).^2;
```

If all the values of **d** are less than 30, the vector **upper** will be an empty vector, and the reference to **d(upper)** and **velocity(upper)** will not cause any values to change.

PRACTICE!

Determine the value of the given expressions. Then check your answers using MAT-LAB. Remember that to check your answers, all you need to do is enter the expression. Assume that the matrix **b** has the following value:

$$b = \begin{bmatrix} 1 & 0 & 4 \\ 0 & 0 & 3 \\ 8 & 7 & 0 \end{bmatrix}$$

a. `any (b)`
b. `find (b)`
c. `all (any(b))`

d. `any(all(b))`
e. `finite(b(:,3))`
f. `any(b(1:2,1:3))`

3.5.6   Loops

A loop is structure that allows you to repeat a set of statements. In general, you should avoid loops in MATLAB, because they are seldom needed and they can significantly increase the execution time of a program. However, there are occasions when loops are needed, so we give a brief introduction to **for** loops and **while** loops.

In the example at the end of the previous discussion, we used the **find** function to find distance values greater than 30 feet and distance values less than or equal to 30 feet. We then computed the corresponding velocites using the appropriate equations. Another way to perform these steps uses a **for** loop. In the statements shown next, the value of **k** is set to 1, and the statements inside the loop are executed. The value of **k** is incremented to 2, and the statements inside the loop are executed again. This process continues until the value of **k** is greater than the length of the vector **d**.

```
for k = 1:length(d)
    if d(k) < 30
            velocity(k) = 0.425 - 0.00175*d(k).^2;
    else
            velocity(k) = 0.625 + 0.12*d(k) ...
                        -0.00025*d(k).^2;
    end
end
```

While these statements perform the same operations as the steps using the **find** function, the solution without a loop will execute much faster.

A **for** loop has the following general structure:

```
for index = expression
    statements
end
```

The expression is a matrix (which could be a scalar or a vector), and the statements are repeated as many times as there are columns in the expression matrix. Each time through the loop, the index has the value of one of the elements in the expression matrix. The rules for writing and using a **for** loop are the following:

a.   The index of a **for** loop must be a variable.

b. If the expression matrix is the empty matrix, the loop will not be executed. Control will pass to the statement following the **end** statement.

c. If the expression matrix is a scalar, the loop will be executed one time, with the index containing the value of the scalar.

d. If the expression matrix is a row vector, each time through the loop, the index will contain the next value in the vector.

e. If the expression matrix is a matrix, each time through the loop, the index will contain the next column in the matrix.

f. Upon completion of a **for** loop, the index contains the last value used.

g. The colon operator can be used to define the expression matrix, using the following format:

```
for k = initial:increment:limit
```

PRACTICE!

Determine the number of times that the given **for** loops will be executed. Then, to check your answers, use the **length** function, which returns the number of values in a vector. Thus, the number of times that the **for** loop in Problem 1 is executed is `length(3:20)`.

a. `for k = 3:20`
b. `for count = -2:14`
c. `for k = -2:-1:-10`
d. `for time = 10: -1:0`
e. `for time = 10:5`
f. `for index = 2:3:12`

The **while** loop is a structure for repeating a set of statements as long as a specified condition is true. The general format for this control structure is

```
while expression
    statements
end
```

The statements in the **while** loop are executed as long as the real part of the **expression** has all nonzero elements. The expression is retested at the end of every loop. If the expression still contains all nonzero elements—that is, it is still true—the group of statements is executed again. When the expression is evaluated as false, control skips to the statement following the **end** statement.

The variables modified in the statements should include the variables in the expression, or else the value of the expression will never change. If the expression is always true, then the loop will execute an infinite number of times (or until you stop the program). The **break** statement can be used to terminate a loop prematurely. A **break** statement will cause termination of the smallest enclosing **while** or **for** loop.

EXAMPLE 3.7: This example shows you how to use the **break** statement to exit a **while** loop:

```
fprintf('Enter a number to find its square root. Enter 0 to
  quit.\n');
while 1
  n = input('Enter a number: ');
  if ~n
    break;
  end
  fprintf('N = %f, Its square root = %f\n', n, sqrt(n));
end;
fprintf('Thank you. All done!\n\n');
```

3.6 SCRIPTS AND USER-WRITTEN FUNCTIONS

In addition to providing an interactive computational environment, MATLAB contains a powerful programming language. As a programmer, you can create and save code in files. The MATLAB files that contain programming code are called M-files. An M-file is an ASCII text file similar to a C or Fortran source code file. An M-file can be created and edited using the MATLAB M-file Editor/Debugger, or you can use another text editor of your choice.

There are two types of M-files, called **scripts** and **functions**. A script is simply a list of MATLAB statements that are saved in a file (typically with a **.m** extension). The script has access to the workspace variables. Any variables created in the script are accessible to the workspace when the script finishes. A script can be executed by typing the filename or by using the **run** command. The run command has two forms, one that accepts modifiers as arguments and one that uses function notation.

EXAMPLE 3.8: Assume that you have created a script file named **myscript.m.** This example shows you three ways of executing the script. In each case, the **.m** file extension is assumed.

```
myscript        - type the file name
run myscript    - use the run command with a filename argument
run ('myscript')- use the functional form of the run command
```

The second type of M-file is called a **function**. A **function** is a special type of script that can accept input arguments and return output arguments. Variables declared within the function are local to the function by default. You can explicitly declare variables as global. The function file has very specific rules and a format that must be followed when writing it. Before we list the rules, we consider a simple example.

EXAMPLE 3.9: The sinc function is a function commonly used in many engineering applications. Unfortunately, there are two widely accepted definitions for this function, shown as follows:

$$f_1(x) = \frac{\sin \pi x}{\pi x} \quad f_2(x) = \frac{\sin x}{x}$$

Both of these functions have an indeterminate form of 0/0 when x is equal to zero. In this case, l'Hôpital's theorem from calculus can be used to prove that both functions are equal to 1 when x is equal to zero. For values of x not equal to zero, these two functions have a similar form. The first function, $f_1(x)$, crosses the x-axis when x is an integer; the second function, $f_2(x)$, crosses the x-axis when x is a multiple of π.

A MATLAB **sinc** function that uses the first definition is included in the Signal Processing Toolbox and in the Student Edition Version 6. Assume that you would like to define another function called **sinc_x** that uses the second definition, which is shown graphically in Figure 3.5. The following function can be used to compute this alternative form of the sinc function, where x can be a scalar, vector, or matrix:

```
function s = sinc_x(x)
%
% SINC_X This function computes the value of sin(x)/x
%
s = x;
set1 = find(abs(x) < 0.001);
set2 = find(abs(x) >= 0.001);
s(set1) = ones (size(set1);
s(set2) = sin(x(set2))./x(set2);
```

Save this function definition in a file named **sinc_x.m**. Now MATLAB programs and scripts can refer to this function in same way that they refer to functions such as **sqrt** and **abs**. The figure shown in Figure 3.5 was plotted using the following statements:

```
% Statements used to generate Figure 3.5.
%
x = -15:0.1:15;
y = sinc_x(x);
plot(x,y),title('Sinc Function'),
xlabel('x'),ylabel('y'),grid,pause
```

We now summarize the rules for writing an M-file function. Refer to the **sinc_x** function as you read each rule.

    a.   The function must begin with a line containing the word **function**, which is followed by the output argument, an equals sign, and the name of the function. The input arguments to the function follow the name of the function and are enclosed in parentheses. This line distinguishes the function file from a script file.

    b.   The first few lines of the function should be comments, because they will be displayed if help is requested for the function name, as in **help sinc_x**.

    c.   The only information returned from the function is contained in the output arguments, which are, of course, matrices. Always check to be sure that the function includes a statement that assigns a value to the output argument.

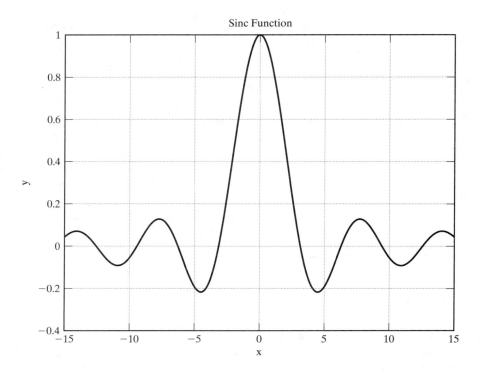

Figure 3.5. Sinc function in interval [−15, 15].

d. The same matrix names can be used in both a function and the program that references it. No confusion occurs as to which matrix is referenced, because the function and the program are completely separate. However, any values computed in the function, other than the output arguments, are not accessible from the program.

e. A function that is going to return more than one value should show all values to be returned as a vector in the function statement, as in the following example, which will return three values:

```
function [dist, vel, accel] = motion(x)
```

All three values need to be computed within the function.

f. A function that has multiple input arguments must list the arguments in the function statement, as shown in the following example, which has two input arguments:

```
function error = mse(w,d)
```

g. The special variables **nargin** and **nargout** can be used to determine the number of input arguments and the number of output arguments for a function.

The **what** command lists all the M-files and MAT files that are available in the current workspace. The **type** command followed by a filename will display the contents of a file on the screen. If an extension is not included with the filename, the **type** command automatically assumes that the extension is **.m**.

SUMMARY

In this chapter, we explored the various MATLAB functions for creating matrices and for calculating new matrices from existing matrices. These functions included mathematical functions, trigonometric functions, data analysis functions, random-number generation functions, and logical functions. In addition, we presented selection statements and functions, so that we can analyze or modify selected portions of matrices. A brief discussion of loops was also included, because they are occasionally needed in MATLAB solutions. Finally, we illustrated the steps in developing a user-written function.

MATLAB SUMMARY

This MATLAB summary lists and briefly describes all of the special characters, commands, and functions that were defined in this chapter.

Special Characters

<	less than
<=	less than or equal to
>	greater than
>=	greater than or equal to
= =	equal to
~=	not equal to
&	and
|	or
~	not

Commands and Functions

abs	computes absolute value or magnitude
acos	computes arccosine
all	determines whether all values are true
any	determines whether any values are true
asin	computes arcsine
atan	computes two-quadrant arctangent
atan2	computes four-quadrant arctangent
break	breaks out of the innermost **for** or **while** loop
ceil	rounds towards ∞
cos	computes cosine
cumprod	determines cumulative products
cumsum	determines cumulative sums
else	optional clause in **if** structure
elseif	optional clause in **if** structure
end	defines end of a control structure
etime	returns the time in seconds that has elapsed between two vectors
exp	computes value with base e

Commands and Functions

find	locates nonzero values
finite	determines whether values are finite
fix	rounds toward zero
floor	rounds toward $-\infty$
for	generates loop structure
function	generates user-defined function
hist	plots histogram
if	tests logical expression
isempty	determines whether matrix is empty
isnan	determines whether values are NaNs
length	determines number of values in a vector
log	computes natural logarithm
log10	computes common logarithm
max	determines maximum value
mean	determines mean value
median	determines median value
min	determines minimum value
prod	determines product of values
rand	generates a uniform random number
randn	generates a Gaussian random number
rem	computes remainder from division
round	rounds to nearest integer
sign	generates $-1, 0, 1$ based on sign
sin	computes sine
sort	sorts values
sqrt	computes square root
std	computes standard deviation
sum	determines sum of values
tan	computes tangent of angle
what	lists variables
while	generates a loop structure

KEY TERMS

else	logical condition	scripts
else if	loops	selection statements
end	mean	simulation
for	median	standard deviation
function	Monte Carlo simulation	uniform random numbers
functions	normal random variable	variance
histogram	not	while
if	or	xor

Problems

1. The range of an object shot at an angle θ with respect to the x-axis and an initial velocity v_0 is given by

$$R(\theta) = \frac{v^2}{g}\sin(2\theta) \text{ for } 0 \le \theta \le \frac{\pi}{2} \text{ and neglecting air resistance.}$$

   Use $g = 9.8$ m/s^2 and an initial velocity of 100 m/s. Show that the maximum range is obtained at $\theta = \frac{\pi}{4}$ by computing the range in increments of 0.05 from $0 \le \theta \le \frac{\pi}{2}$.

2. The vector **G** represents the distribution of final grades in a statics course. Compute the mean, median, and standard deviation of **G** and plot a histogram of the distribution. Which better represents the "most typical" grade, the mean or the median? Why?

   $\mathbf{G} = [68, 83, 61, 70, 75, 82, 57, 5, 76, 85, 62, 71, 96, 78, 76, 68, 72, 75, 83, 93]$

3. Generate 10,000 Gaussian random numbers with a mean of 80.0 and standard deviation of 23.5. Plot a histogram of the random numbers using 50 bins. Does the distribution appear to be a normal distribution?

4. In general, using a **for** loop is less efficient in MATLAB than using array operations. Test this assertion by timing a long array multiplication. Create matrix **A** containing 10,000 ones. Compare the results of multiplying each element by π using a **for** loop and using array multiplication. You can time the results by using the **clock** function and the function **etime**, which measures elapsed time.

   ```
   t0 = clock;
   ...
   code to be timed
   ...
   etime (clock, t0)
   ```

5. Create a user-defined function named **grades** that takes a vector of grades as an input argument and performs the requirements of Problem 2. You can use your function to solve Problem 2 by entering:

   ```
   grades (G);
   ```

6. **Rocket Trajectory.** A small rocket is being designed to make wind shear measurements in the vicinity of thunderstorms. Before testing begins, the designers are developing a simulation of the rocket's trajectory. They have derived the following equation, which they believe will predict the performance of the test rocket, where t is the elapsed time, in seconds:

   $$\text{height} = 60 + 2.13t^2 - 0.0013t^4 + 0.000034t^{4.751}$$

   The equation gives the height of the rocket above ground level at time t. The first term (60) is the height of the nose of the rocket above ground level, in feet.

   - Give the commands to compute and print the time and height of the rocket from the time that it hits the ground in increments of 2 seconds. If the rocket has not hit the ground within 100 seconds, print values only up through 100 seconds.

• Modify the steps in part (a) so that instead of a table, the program prints the time at which the rocket begins falling back to the ground and the time at which the rocket impacts the ground.

7. **Suture Packaging.** Sutures are strands or fibers used to sew living tissue together after an injury or an operation. Packages of sutures must be sealed carefully before they are shipped to hospitals, so that contaminants cannot enter the packages. The object that seals the package is referred to as the *sealing die*. Generally, sealing dies are heated with an electric heater. For the sealing process to be a success, the sealing die is maintained at an established temperature and must contact the package with a predetermined pressure for an established period of time. The period of time in which the sealing die contacts the package is called the *dwell time*. Assume that the range of parameters for an acceptable seal are the following:

Temperature: 150–170°

Pressure: 60–70 psi

Dwell Time: 2.0–2.5 s

• A data file named **suture.dat** contains information on batches of sutures that have been rejected during a one-week period. Each line in the data file contains the batch number, the temperature, the pressure, and the dwell time for a rejected batch. A quality-control engineer would like to analyze this information to determine the percent of the batches rejected due to temperature, the percent rejected due to pressure, and the percent rejected due to dwell time. If a specific batch is rejected for more than one reason, it should be counted in all applicable totals. Give the MATLAB statements to compute and print these three percentages. Use the following data:

BATCH NUMBER	TEMPERATURE	PRESSURE	DWELL TIME
24551	145.5	62.3	2.23
24582	153.7	63.2	2.52
26553	160.3	58.9	2.51
26623	159.5	58.9	2.01
26642	160.3	61.2	1.98

• Modify the solution developed in part (a) so that it also prints the number of batches in each rejection category and the total number of batches rejected. (Remember that a rejected batch should appear only once in the total, but could appear in more than one rejection category.)

• Write a program to read the data file **suture.dat**, and make sure that the information relates only to batches that should have been rejected. If any batch should not be in the data file, print an appropriate message with the batch information.

8. **Timber Regrowth.** A problem in timber management is to determine how much of an area to leave uncut so that the harvested area is reforested in a certain period of time. It is assumed that reforestation takes place at a known rate per year, depending on climate and soil conditions. A reforestation equation expresses this growth as a function of the amount of timber standing and the reforestation rate. For example, if 100 acres are left standing after harvesting and the reforestation rate is 0.05, then 105 acres are forested at the end of the first year. At the end of the second year, the number of acres forested is 110.25 acres.

- Assume that there are 14,000 acres total, with 2,500 uncut acres, and that the reforestation rate is 0.02. Print a table showing the number of acres reforested at the end of each year, for a total of 20 years.
- Modify the program developed in part (a) so that the user can enter the number of years to be used for the table.
- Modify the program developed in part (a) so that the user can enter a number of acres, and the program will determine how many years are required for the number of acres to be forested.

9. **Sensor Data.** Suppose that a data file named **sensor.dat** contains information collected from a set of sensors. Each row contains a set of sensor readings, with the first row containing values collected at 0.0 seconds, the second row containing values collected at 1.0 seconds, and so on.

- Write a program to read the data file and print the number of sensors and the number of seconds of data contained in the file.
- Write a program to preprocess the sensor data so that all values greater than 10.0 are set to 10.0 and all values less than −10.0 are set to −10.0.
- Write a program to print the subscripts of sensor data values with an absolute value greater than 20.0.
- Write a program to print the percentage of sensor data values that are zero.

10. **Power Plant Output.** The power output in megawatts from a power plant over a period of eight weeks has been stored in a data file named **plant.dat**. Each line in the data file represents data for one week and contains the output for day 1, day 2, . . . , day 7.

- Write a program that uses the power plant output data and prints a report that lists the number of days with greater-than-average power output. The report should give the week number and the day number for each of these days, in addition to printing the average power output for the plant during the eight-week period.
- Write a program that uses the power plant output data and prints the day and week during which the maximum and minimum power output occurred. If the maximum or minimum power output occurred on more than one day, the program should print all the days involved.
- Write a program that uses the power plant output data to print the average power output for each week. Also, print the average power output for day 1, day 2, and so on.

11. **Faster Loops.** When possible it is better to avoid using **for** loops, because they are slow to execute. Generate a 100,000-item vector of random digits, square this vector as an array, and use the commands **tic** and **toc** to time the operation. Next, perform the same operation element by element in a **for** loop. Again, time the operation using **tic** and **toc**. Next, convince yourself that suppressing the printing of intermediate answers will speed execution of the code by allowing these same operations to run and print the answers as they are calculated. If you are going to be using a constant value several times in a **for** loop, calculate it once and store it, rather than calculating it each time through the loop. Demonstrate the speed increase of this process by adding sin(0.3) to every value in the long vector in a **for** loop. If MATLAB must increase the size of a vector every time through a loop, the process will take more time than if the vector were already the appropriate size. Demonstrate this fact by storing the value of the foregoing addition into a new vector. Speed execution of the code by making the vector that will store the values equal to a zero vector of the appropriate size before the loop is entered.

12. **Your Toolbox.** You will eventually wish to create a directory of functions that you have created for MATLAB that will be used in many of the programs you write. The first two functions that you might wish to add are **RD(x)** and **DR(x)**, which convert from radians to degrees and back, respectively. These functions will be useful, because it is often easier for people to think in degrees, but MATLAB computes in radians. The functions will make your code cleaner by making the conversion easy. The two functions should be in their own function files and should be put into a new directory that is pathed using the path dialog in MATLAB, so that every time you start MATLAB, the functions will be available. You should put helpful comments in the functions, so that when you ask for help on the function, relevant text is available. As you add more functions to this directory, you may wish to consider adding a **contents.m** file that would serve as a help file for your entire personal toolbox.

4

Matrix Computations

GRAND CHALLENGE: MAPPING THE HUMAN GENOME

Some of the goals of the Human Genome Project are to

- *identify* all the genes in human DNA,
- *determine* the sequences of the chemical base pairs that make up human DNA,
- *store* this information in databases,
- *improve* tools for data analysis,
- *transfer* related technologies to the private sector, and
- *address* the ethical, legal, and social issues that may arise from the project.

The majority of the human genome was sequenced in the year 2000, and 90 percent of the sequence of the genome's three billion base-pairs was published February 2001. One of the surprises from the results was the smaller-than-expected number of genes in the human genome. Early estimates of the number of genes ranged from 50,000 to 100,000. The deciphering of the human genome resulted in mapping roughly 30,000 genes. Another interesting result occurred when the human genome was matched to a mouse genome. When scientists compared the human and mouse

SECTIONS

4.1 Matrix Operations and Functions
4.2 Solutions to Systems of Linear Equations

OBJECTIVES

After reading this chapter, you should be able to

- perform operations that apply to an entire matrix as a unit and
- solve simultaneous equations using MATLAB.

genomes, they discovered that more than 90 percent of the mouse genome could be lined up with a region on the human genome.

The U.S. Department of Energy sponsors a comprehensive website containing information about the Human Genome Project at

http://www.ornl.gov/hgmis/

4.1 MATRIX OPERATIONS AND FUNCTIONS

Many engineering computations use a matrix as a convenient way to represent a set of data. In this chapter, we are generally concerned with matrices that have more than one row and more than one column. Recall that scalar multiplication and matrix addition and subtraction are performed element by element. Matrix multiplication is covered in this section. Matrix division is presented in the next section and is used to compute the solution to a set of simultaneous linear equations.

4.1.1 Transpose

The **transpose** of a matrix is a new matrix in which the rows of the original matrix are the columns of the new matrix. We use a superscript **T** after the name of a matrix to refer to the transpose of the matrix. For example, consider the following matrix and its transpose:

$$\mathbf{A} = \begin{bmatrix} 2 & 5 & 1 \\ 7 & 3 & 8 \\ 4 & 5 & 21 \\ 16 & 13 & 0 \end{bmatrix} \quad \mathbf{A}^{\mathrm{T}} = \begin{bmatrix} 2 & 7 & 4 & 16 \\ 5 & 3 & 5 & 13 \\ 1 & 8 & 21 & 0 \end{bmatrix}$$

If we consider a couple of the elements, we see that the value in position (3,1) of **A** has now moved to position (1,3) of $\mathbf{A}^{\mathrm{T}}$, and the value in position (4,2) of **A** has now moved to position (2,4) of $\mathbf{A}^{\mathrm{T}}$. In general, the row and column subscripts are interchanged to form the transpose; hence, the value in position (i,j) is moved to position (j,i).

In MATLAB, the transpose of the matrix **A** is denoted by **A'**. Observe that the transpose will have a different size than the original matrix if the original matrix is not a square matrix. We frequently use the transpose operation to convert a row vector to a column vector or a column vector to a row vector.

The **dot product** is a scalar computed from two vectors of the same size. This scalar is the sum of the products of the values in corresponding positions in the vectors, as shown in the following summation equation, which assumes that there are n elements in the vectors **A** and **B**:

$$\text{dot product} = \mathbf{A} \cdot \mathbf{B} = \sum_{i=1}^{n} a_i b_i$$

In MATLAB, we can compute the dot product with the following statement:

```
dot_product = sum(A.*B);
```

Recall that **A.*B** contains the results of an elementwise multiplication of **A** and **B**. When **A** and **B** are both row vectors or are both column vectors, **A.*B** is also a vector. We then sum the elements in this vector, thus yielding the dot product. The **dot** function may also be used to compute the dot product:

```
dot(A,B);
```

EXAMPLE 4.1: To illustrate, assume that **A** and **B** are the following vectors:

$$\mathbf{A} = [4 \; -1 \; 3] \quad \mathbf{B} = [-2 \; 5 \; 2]$$

The dot product is then

$$\begin{aligned} \mathbf{A} \cdot \mathbf{B} &= 4 \cdot (-2) + (-1) \cdot 5 + 3 \cdot 2 \\ &= (-8) + (-5) + 6 \\ &= -7 \end{aligned}$$

You can test this result by typing

```
dot(A,B);
```

4.1.2   Matrix Multiplication

Matrix multiplication is not accomplished by multiplying corresponding elements of the matrices. In matrix multiplication, the value in position $c(i,j)$ of the product **C** of two matrices **A** and **B** is the dot product of row i of the first matrix and column j of the second matrix, as shown in the following summation equation:

$$c_{i,j} = \sum_{k=1}^{N} a_{ik} b_{kj}$$

Because the dot product requires that the vectors have the same number of elements, the first matrix (**A**) must have the same number of elements (N) in each row as there are in each column of the second matrix (**B**). Thus, if **A** and **B** both have five rows and five columns, their product has five rows and five columns. Furthermore, for these matrices, we can compute both **AB** and **BA**, but in general, they will not be equal.

If **A** has two rows and three columns and **B** has three rows and three columns, the product **AB** will have two rows and three columns. To illustrate, consider the following matrices:

$$\mathbf{A} = \begin{bmatrix} 2 & 5 & 1 \\ 0 & 3 & -1 \end{bmatrix} \quad \mathbf{B} = \begin{bmatrix} 1 & 0 & 2 \\ -1 & 4 & -2 \\ 5 & 2 & 1 \end{bmatrix}$$

The first element in the product **C** = **AB** is

$$\begin{aligned} c_{1,1} &= \sum_{k=1}^{3} a_{1k} b_{k1} \\ &= a_{1,1} b_{1,1} + a_{1,2} b_{2,1} + a_{1,3} b_{3,1} \\ &= 2 \cdot 1 + 5 \cdot (-1) + 1 \cdot 5 \\ &= 2 \end{aligned}$$

Similarly, we can compute the rest of the elements in the product of **A** and **B**:

$$\mathbf{AB} = \mathbf{C} = \begin{bmatrix} 2 & 22 & -5 \\ -8 & 10 & -7 \end{bmatrix}$$

In this example, we cannot compute **BA**, because **B** does not have the same number of elements in each row as **A** has in each column.

An easy way to decide if a matrix product exists is to write the sizes of the two matrices side by side. Then, if the two inside numbers are the same, the product exists, and the size of the product is determined by the two outside numbers. To illustrate, in the previous example, the size of **A** is 2×3, and the size of **B** is 3×3. Therefore, if we want to compute **AB**, we write the sizes side by side:

$$2 \times 3, \ 3 \times 3$$

The two inner numbers are both the value 3, so **AB** exists, and its size is determined by the two outer numbers, 2×3. If we want to compute **BA**, we again write the sizes side by side:

$$3 \times 3, \ 2 \times 3$$

The two inner numbers are not the same, so **BA** does not exist. If the two inner numbers are the same, then **A** is said to be **conformable for multiplication** to **B**.

In MATLAB, matrix multiplication is denoted by an asterisk. Thus, the command to perform matrix multiplication of matrices **A** and **B** is

```
A * B;
```

EXAMPLE 4.2:

Generate the matrices in our previous example, and then compute the matrix product:

```
A = [2,5,1; 0,3,-1 ];
B = [1,0,2; -1,4,-2; 5,2,1 ];
C = A * B;
```

The results are as follows:

```
C =
        2      22      -5
       -8      10      -7
```

Note that **B** * **A** does not exist, because the number of columns of **B** does not equal the number of rows of **A**. In other words, **B** is not conformable for multiplication with **A**. Execute the MATLAB command **C** = **B** * **A**:

```
C = B*A;
```

You will get the following warning message:

```
??? Error using ==> *
Inner matrix dimensions must agree.
```

Assume that **I** is a square **identity matrix**. (Recall that an identity matrix is a matrix with ones on the main diagonal and zeros elsewhere.) If **A** is a square matrix of the same size, then **A** * **I** and **I** * **A** are both equal to **A**.

EXAMPLE 4.3:   Generate a square identity matrix **I** of order 3×3:

```
I = eye(3);
```

Use the square matrix **A** from the preceding discussion to verify that $\mathbf{A} * \mathbf{I} = \mathbf{I} * \mathbf{A}$:

```
A * I == I * A % 1 if true, 0 if false
C = A * I;
C =
        1       0       2
       -1       4      -2
        5       2       1
C = I * A;
C =
        1       0       2
       -1       4      -2
        5       2       1
```

4.1.3   Matrix Powers

Recall that if **A** is a matrix, then the operation A.^2 squares each element in **A**. If we want to square the matrix—that is, to compute $\mathbf{A}*\mathbf{A}$—we use the operation A^2. A^4 is equivalent to $\mathbf{A}*\mathbf{A}*\mathbf{A}*\mathbf{A}$. To perform a matrix multiplication between two matrices, the number of rows in the first matrix must be the same value as the number of columns in the second matrix. Therefore, to raise a matrix to a power, the number of rows must equal the number of columns, and thus the matrix must be a square matrix.

4.1.4   Matrix Inverse

By definition, the **inverse** of a square matrix **A** is the matrix $\mathbf{A}^{-1}$ such that the matrix products $\mathbf{A}\mathbf{A}^{-1}$ and $\mathbf{A}^{-1}\mathbf{A}$ are both equal to the identity matrix. For example, consider the following two matrices **A** and **B**:

$$\mathbf{A} = \begin{bmatrix} 2 & 1 \\ 4 & 3 \end{bmatrix} \qquad \mathbf{B} = \begin{bmatrix} 1.5 & -0.5 \\ -2 & 1 \end{bmatrix}$$

If we compute the products **AB** and **BA**, we obtain the following matrices (do the matrix multiplications by hand to be sure you follow the steps):

$$\mathbf{AB} = \begin{bmatrix} 1 & 0 \\ 0 & 1 \end{bmatrix} \qquad \mathbf{BA} = \begin{bmatrix} 1 & 0 \\ 0 & 1 \end{bmatrix}$$

Therefore, **A** and **B** are inverses of each other, or $\mathbf{A} = \mathbf{B}^{-1}$ and $\mathbf{B} = \mathbf{A}^{-1}$.

Computing the inverse of a matrix is a tedious process; fortunately, MATLAB contains an **inv** function that performs the computations for us. (We do not present the steps for computing an inverse in this text. Refer to a linear algebra text if you are interested in the techniques for computing an inverse.) Thus, if we execute **inv(A)**, using the matrix **A** defined previously the result will be the matrix **B**. Similarly, if we execute **inv(B)**, the result should be the matrix **A**. Try this yourself.

There are matrices for which an inverse does not exist; these matrices are called **singular**, or **ill-conditioned matrices**. When you attempt to compute the inverse of an ill-conditioned matrix in MATLAB, an error message is printed.

EXAMPLE 4.4:   Create the matrix $\mathbf{A} = [\ 1, 2; 3, 4\ ]$.
Raise the matrix to the second power using the following command:

```
C = A^2;
```

The results will be as follows:

```
C =
        7      10
       15      22
```

Note that raising $\mathbf{A}$ to the **matrix power** of two is different from raising $\mathbf{A}$ to the **array power** of two:

```
C = A.^2;
```

Raising $\mathbf{A}$ to the array power of two produces the following results:

```
C =
        1       4
        9      16
```

4.1.5   Determinants

A **determinant** is a scalar computed from the entries in a square matrix. Determinants have various applications in engineering, including computing inverses and solving systems of simultaneous equations. For a 2×2 matrix $\mathbf{A}$, the determinant is

$$\left|\mathbf{A}\right| = a_{1,1}a_{2,2} - a_{2,1}a_{1,2}$$

Therefore, the determinant of $\mathbf{A}$, or $\left|\mathbf{A}\right|$, is equal to 8 for the following matrix:

$$\mathbf{A} = \begin{bmatrix} 1 & 3 \\ -1 & 5 \end{bmatrix}$$

For a 3×3 matrix $\mathbf{A}$, the determinant is the following:

$$\left|\mathbf{A}\right| = a_{1,1}a_{2,2}a_{3,3} + a_{1,2}a_{2,3}a_{3,1} + a_{1,3}a_{2,1}a_{3,2} - a_{3,1}a_{2,2}a_{1,3} - a_{3,2}a_{2,3}a_{1,1} - a_{3,3}a_{2,1}a_{1,2}$$

If

$$\mathbf{A} = \begin{bmatrix} 1 & 3 & 0 \\ -1 & 5 & 2 \\ 1 & 2 & 1 \end{bmatrix}$$

then $\left|\mathbf{A}\right|$ is equal to $5 + 6 + 0 - 0 - 4 - (-3)$, or 10.

A more involved process is necessary for computing determinants of matrices with more than three rows and columns. We do not include a discussion of the process for computing a general determinant here, because MATLAB will automatically compute a determinant using the **det** function, with a square matrix as its argument, as in **det(A)**.

PRACTICE!

Use MATLAB to define the following matrices:

$$A = \begin{bmatrix} 2 & 1 \\ 0 & -1 \\ 3 & 0 \end{bmatrix} \quad B = \begin{bmatrix} 1 & 3 \\ -1 & 5 \end{bmatrix}$$

$$C = \begin{bmatrix} 3 & 2 \\ -1 & -2 \\ 0 & 2 \end{bmatrix} \quad D = \begin{bmatrix} 1 & 2 \end{bmatrix} \quad I = \begin{bmatrix} 1 & 0 \\ 0 & 1 \end{bmatrix}$$

Compute the matrix specified in each problems if the matrix exists.

  a. **DB**
  b. **BC**^{T}
  c. **(CB)D**^{T}
  d. **B**$^{-1}$**B**
  e. **AC**^{T}
  f. **(AC**^{T}**)**$^{-1}$
  g. **det(B)**
  h. **det(A*C**^{T}**)**

4.2   SOLUTIONS TO SYSTEMS OF LINEAR EQUATIONS

Consider the following system of three equations with three unknowns:

$$\begin{array}{rrrcr} 3x & +2y & -z & = & 10 \\ -x & +3y & +2z & = & 5 \\ x & -y & -z & = & -1 \end{array}$$

We can rewrite this system of equations using the following matrices:

$$A = \begin{bmatrix} 3 & 2 & -1 \\ -1 & 3 & 2 \\ 1 & -1 & -1 \end{bmatrix} \quad X = \begin{bmatrix} x \\ y \\ z \end{bmatrix} \quad B = \begin{bmatrix} 10 \\ 5 \\ -1 \end{bmatrix}$$

Using matrix multiplication, the **system of equations** can then be written as $AX = B$. Go through the multiplication to convince yourself that this matrix equation yields the original set of equations.

To simplify the notation, we designate the variables as x_1, x_2, x_3, and so on. Rewriting the initial set of equations using this notation, we have

$$\begin{array}{rrrcr} 3x_1 & +2x_2 & -x_3 & = & 10 \\ -x_1 & +3x_2 & +2x_3 & = & 5 \\ x_1 & -x_2 & -x_3 & = & -1 \end{array}$$

This set of equations is then represented by the matrix equation $\mathbf{AX} = \mathbf{B}$, where $\mathbf{X}$ is the column vector $[x_1, x_2, x_3]^\mathsf{T}$. We now present two methods for solving a system of N equations with N unknowns.

4.2.1 Solution Using the Matrix Inverse

One way to solve a system of equations is by using the matrix inverse. For example, assume that $\mathbf{A}$, $\mathbf{X}$, and $\mathbf{B}$ are the matrices defined earlier in this section:

$$\mathbf{A} = \begin{bmatrix} 3 & 2 & -1 \\ -1 & 3 & 2 \\ 1 & -1 & -1 \end{bmatrix} \quad \mathbf{X} = \begin{bmatrix} x_1 \\ x_2 \\ x_3 \end{bmatrix} \quad \mathbf{B} = \begin{bmatrix} 10 \\ 5 \\ -1 \end{bmatrix}$$

Then $\mathbf{AX} = \mathbf{B}$. If we premultiply both sides of this matrix equation by $\mathbf{A}^{-1}$, we have $\mathbf{A}^{-1}\mathbf{AX} = \mathbf{A}^{-1}\mathbf{B}$. However, because $\mathbf{A}^{-1}\mathbf{A}$ is equal to the identity matrix $\mathbf{I}$, we have $\mathbf{IX} = \mathbf{A}^{-1}\mathbf{B}$, or $\mathbf{X} = \mathbf{A}^{-1}\mathbf{B}$. In MATLAB, we can compute this solution using the following command:

```
X = inv(A)*B;
```

EXAMPLE 4.5:

As an example, we will solve the following system of equations:

$$3x_1 + 5x_2 = -7$$
$$2x_1 - 4x_2 = 10$$

Type the following MATLAB commands to define $\mathbf{A}$ and $\mathbf{B}$:

```
A = [3 5; 2 -4];
B = [-7 10]';
```

Now solve for $\mathbf{X}$ using the inverse of $\mathbf{A}$:

```
X = inv(A)*B;
```

MATLAB finds the following solution:

```
X =
        1.0000
       -2.0000
```

4.2.2 Solution Using Matrix Left Division

A better way to solve a system of linear equations is to use the matrix division operator:

```
X = A\B;
```

This method produces the solution using Gaussian elimination, without forming the inverse. Using the matrix division operator is more efficient than using the matrix inverse and produces a greater numerical accuracy.

EXAMPLE 4.6: As an example, we will solve the same system of equations used in the previous example:

$$3x_1 + 5x_2 = -7$$
$$2x_1 - 4x_2 = 10$$

However, now solve for **X** by using matrix left division:

    X = A\B;

Again, MATLAB finds the following solution:

    X =
         1
        -2

To confirm that the values of **X** do indeed solve each equation, we can multiply **A** by **X** using the expression **A*X**. The result is the column vector $[-7, 10]^T$.

If there is not a unique solution to a system of equations, an error message is displayed. The solution vector may contain values of NaN, ∞, or $-\infty$, depending on the values of the matrices **A** and **B**.

PRACTICE!

Solve the given systems of equations using both matrix left division and inverse matrices. Use MATLAB to verify that each solution solves the system of equations using matrix multiplication.

a.
$$-2x_1 + x_2 = -3$$
$$x_1 + x_2 = 3$$

b.
$$10x_1 - 7x_2 + 0x_3 = 7$$
$$-3x_1 + 2x_2 + 6x_3 = 4$$
$$5x_1 + x_2 + 5x_3 = 6$$

c.
$$x_1 + 4x_2 - x_3 + x_4 = 2$$
$$2x_1 + 7x_2 + x_3 - 2x_4 = 16$$
$$x_1 + 4x_2 - x_3 + 2x_4 = -15$$
$$3x_1 - 10x_2 - 2x_3 + 5x_4 = -15$$

SUMMARY

In this chapter, we defined the transpose, the inverse, and the determinant of a matrix. We also defined the computation of a dot product (between two vectors) and a matrix product (between two matrices). Two methods for solving a system of N equations with N unknowns using matrix operations were presented. One method used matrix left division, and the other used the inverse of a matrix.

MATLAB SUMMARY

This MATLAB summary lists and briefly describes all of the special characters, commands, and functions that were defined in this chapter.

Special Characters	
'	indicates a matrix transpose
*	indicates matrix multiplication
\	indicates matrix left division

Commands and Functions	
det	computes the determinant of a matrix
inv	computes the inverse of a matrix

KEY TERMS

determinant	inverse	transpose
dot product	matrix multiplication	
identity matrix	system of equations	

Problems

1. Compute the dot product of the following pairs of vectors by hand, showing your work, and then give two different MATLAB commands to compute the dot product of each pair:

    a. $\mathbf{A} = [1 \ 3 \ 5]$, $\mathbf{B} = [-3 \ -2 \ 4]$
    b. $\mathbf{A} = [0 \ -1 \ -4 \ -8]$, $\mathbf{B} = [4 \ -2 \ -3 \ 24]$

2. Compute the matrix product $\mathbf{A}^*\mathbf{B}$ of the following pairs of matrices, showing your work:

    a. $\mathbf{A} = [12 \ 4; 3 \ -5]$, $\mathbf{B} = [\ 2 \ 12; 0 \ 0]$
    b. $\mathbf{A} = [1 \ 3 \ 5; 2 \ 4 \ 6]$, $\mathbf{B} = [-2 \ 4; 3 \ 8; 12 \ -2]$

3. Given the array $\mathbf{A} = [-1 \ 3; 4 \ 2]$, raise $\mathbf{A}$ to the second power by array exponentiation. Raise $\mathbf{A}$ to the second power by matrix exponentiation. Compute by hand and show your work. Give the MATLAB commands for matrix and array exponentiation of $\mathbf{A}$.

4. Given the array $\mathbf{A} = [-1 \ 3; 4 \ 2]$, compute the determinant of $\mathbf{A}$ by hand. Show your work. Give the MATLAB command for computing the determinant of $\mathbf{A}$.

5. If $\mathbf{A}$ is conformable to $\mathbf{B}$ for addition, then a theorem states that $(\mathbf{A} + \mathbf{B})^{\mathrm{T}} = \mathbf{A}^{\mathrm{T}} + \mathbf{B}^{\mathrm{T}}$. Use MATLAB to test this theorem on the following matrices:

$$\mathbf{A} = \begin{bmatrix} 2 & 12 & -5 \\ -3 & 0 & -2 \\ 4 & 2 & -1 \end{bmatrix} \quad \mathbf{B} = \begin{bmatrix} 4 & 0 & 12 \\ 2 & 2 & 0 \\ -6 & 3 & 0 \end{bmatrix}$$

6. Given that matrices $\mathbf{A}$, $\mathbf{B}$, and $\mathbf{C}$ are conformable for multiplication, then the associative property holds—i.e., $\mathbf{A}(\mathbf{BC}) = (\mathbf{AB})\mathbf{C}$. Test the associative property using matrices $\mathbf{A}$ and $\mathbf{B}$ from Problem 1, along with matrix $\mathbf{C}$, which is as follows:

$$\mathbf{C} = \begin{bmatrix} 4 \\ -3 \\ 0 \end{bmatrix}$$

7. Recall that not all matrices have an inverse. A matrix is singular (i.e., it doesn't have an inverse) *iff* $|\mathbf{A}| = 0$. Test the following matrices using the determinant function to see if each has an inverse:

$$\mathbf{A} = \begin{bmatrix} 2 & -1 \\ 4 & 5 \end{bmatrix} \quad \mathbf{B} = \begin{bmatrix} 4 & 2 \\ 2 & 1 \end{bmatrix} \quad \mathbf{C} = \begin{bmatrix} 2 & 0 & 0 \\ 1 & 2 & 2 \\ 5 & -4 & 0 \end{bmatrix}$$

If an inverse exists, compute the inverse.

8. Solve the following system of equations using both the matrix left division and the inverse matrix methods:

$$x_1 - x_2 - x_3 - x_4 = 5$$
$$x_1 + 2x_2 + 3x_3 + x_4 = -2$$
$$2x_1 + 2x_3 + 3x_4 = 3$$
$$3x_1 + x_2 + 2x_4 = 1$$

9. Time each method that you used in Problem 4 by using the **clock** function and the **etime** function, which measures elapsed time. Which method is faster?

```
t0 = clock;
...
code to be timed
...
etime(clock,t0)
```

10. **Single-Voltage-Source Electrical Circuit**. This problem relates to a system of equations generated by the electrical circuit shown in Figure 4.1, which contains a single voltage source and five resistors. The following set of equations defines the currents in this circuit:

$$-V_1 + R_2(i_1 - i_2) + R_4(i_1 - i_3) = 0$$
$$R_1 i_2 + R_3(i_2 - i_3) + R_2(i_2 - i_1) = 0$$
$$R_3(i_3 - i_2) + R_5 i_3 + R_4(i_3 - i_1) = 0$$

Compute the currents using resistor values (R_1, R_2, R_3, R_4) and a voltage value (V_1) entered from the keyboard.

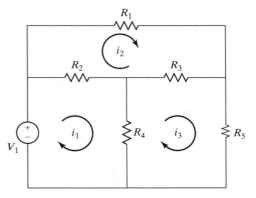

Figure 4.1. Circuit with one voltage source.

11. **Amino Acids**. The amino acids in proteins contain molecules of oxygen (O), carbon (C), nitrogen (N), sulfur (S), and hydrogen (H), as shown in Table 4.1. The molecular weights for oxygen, carbon, nitrogen, sulfur, and hydrogen are as follows:

Oxygen   15.9994
Carbon   12.011
Nitrogen   14.00674
Sulfur   32.066
Hydrogen   1.00794

    a. Write a program in which the user enters the number of oxygen atoms, carbon atoms, nitrogen atoms, sulfur atoms, and hydrogen atoms in an amino acid. Compute and print the corresponding molecular weight. Use a dot product to compute the molecular weight.

    b. Write a program that computes the molecular weight of each amino acid in Table 4.1, assuming that the numeric information in this table is contained in a data file named **elements.dat**. Generate a new data file named **weights.dat** that contains the molecular weights of the amino acids. Use matrix multiplication to compute the molecular weights.

TABLE 4.1   Amino Acid Molecules

AMINO ACID	O	C	N	S	H
Alanine	2	3	1	0	7
Arginine	2	6	4	0	15
Asparagine	3	4	2	0	8
Aspartic	4	4	1	0	6
Cysteine	2	3	1	1	7
Glutamic	4	5	1	0	8
Glutamine	3	5	2	0	10
Glycine	2	2	1	0	5
Histidine	2	6	3	0	10
Isoleucine	2	6	1	0	13
Leucine	2	6	1	0	13
Lysine	2	6	2	0	15
Methionine	2	5	1	1	11
Phenylanlanine	2	9	1	0	11
Proline	2	5	1	0	10
Serine	3	3	1	0	7
Threonine	3	4	1	0	9
Tryptophan	2	11	2	0	11
Tyrosine	3	9	1	9	11
Valine	2	5	1	0	11

c.   Modify the program developed in part (b) so that it also computes and prints the average amino acid molecular weight.

d.   Modify the program developed in part (b) so that it also computes and prints the minimum and maximum molecular weights.

5

Symbolic Mathematics

GRAND CHALLENGE: WEATHER PREDICTION

Weather balloons are used to collect data from the upper atmosphere to use in developing weather models. These balloons are filled with helium and rise to an equilibrium point at which the difference between the density of the helium inside the balloon and the density of the air outside the balloon is just enough to support the weight of the balloon. During the day, the sun warms the balloon, causing it to rise to a new equilibrium point; in the evening, the balloon cools, and it descends to a lower altitude. The balloon can be used to measure the temperature, pressure, humidity, chemical concentrations, or other properties of the air around the balloon. A weather balloon may stay aloft for only a few hours or as long as several years, collecting environmental data. The balloon falls back to earth as the helium leaks out or is released.

SECTIONS

OBJECTIVES

After reading this chapter, you should be able to

- factor and simplify mathematical expressions
- solve systems of equations, and
- determine the symbolic derivative of an expression and integrate an expression.

5.1   SYMBOLIC ALGEBRA

In the previous chapters, we used numbers to make computations in MATLAB; in this chapter, we use symbols to make computations in MATLAB. This capability to manipulate mathematical expressions without using numbers can be very useful in solving certain types of engineering problems. The symbolic functions in MATLAB are based on the **Maple V** software package, which comes from Waterloo Maple, Inc., in Canada. A complete set of these symbolic functions is available in the Symbolic Math Toolbox, which is available for the professional version of MATLAB; a subset of the symbolic functions is included with the student edition of MATLAB, version 6.

In this chapter we focus on **symbolic algebra**, which is used to factor and simplify mathematical expressions, to determine solutions to equations, and to perform integration and differentiation of mathematical expressions. Additional capabilities that we do not discuss in this chapter include linear algebra functions for computing inverses, determinants, eigenvalues, and canonical forms of symbolic matrices; variable precision arithmetic for numerically evaluating mathematical expressions to any specified accuracy; symbolic and numerical solutions to differential equations; and special mathematical functions that evaluate functions such as Fourier transforms. For more details on these additional symbolic capabilities, refer to the second edition of *Engineering Problem Solving with MATLAB*, by Delores M. Etter (Prentice Hall, 1997), or to the Symbolic Math Toolbox documentation.

5.1.1   Symbolic Expressions

A **symbolic expression** is stored in MATLAB as a character string. Single quotes are used to define a symbolic expression. Symbolic objects are created using the **sym** function.

EXAMPLE 5.1:

To create a symbolic object, use the **sym** function with a single argument. The argument should consist of a character string that describes a symbolic expression. As an example, we create a symbolic object **S** for the following expression:

$$x^3 - 2y^2 + 3a$$

To create the symbolic object, type the following command:

```
S = sym('x^3-2*y^2+3*a');
```

To identify the variables in a symbolic expression or matrix, use the **findsym** function. The **findsym** function returns the variables of its single argument in alphabetical order, separated by commas.

EXAMPLE 5.2:

To identify the variables in the symbolic expression **S**, type

```
findsym(S)
```

A list of the variables in **S** is returned:

```
ans =
     a, x, y
```

To substitute variables or expressions within a symbolic expression, use the **subs** function. When used with three arguments, the syntax of the **subs** function is as follows:

```
subs(S,old,new);
```

S is a symbolic expression. **Old** is a symbolic variable or string that represents a variable name. **New** is a symbolic or numeric variable or expression. Multiple substitutions may be made by listing arguments within curly braces.

EXAMPLE 5.3:   To substitute the variable b for variable a in the symbolic expression **S** from the previous example, type the following:

```
subs(S, 'a', 'b')
```

The results are

```
ans =
      x^3-2*y^2+3*(b)
```

To substitute the variables q and r for the variables x and y, respectively, type the following:

```
subs(S,{'x','y'},{'q','r'})
```

The results are:

```
ans =
      q^3-2*r^2+3*a
```

Note that neither of these commands stored the value of the new expression into **S**. For this reason the first substitution was not reflected in the final answer after the second substitution.

PRACTICE!

Create symbolic objects **S1**, **S2**, **S3**, **S4**, and **S5** for the following mathematical expressions, using the **sym** function:

$$S1 = x^2 - 9$$

$$S2 = (x - 3)^2$$

$$S3 = \frac{x^2 - 3x - 10}{x + 2}$$

$$S4 = x^3 + 3x^2 - 13x - 15$$

$$S5 = 2x - 3y + 4x + 13b - 8y$$

MATLAB includes a function called **ezplot** that generates a plot of a symbolic expression of one variable. The independent variable ranges by default over the interval $[-2\pi, 2\pi]$. A second form of **ezplot** allows the user to specify the range. If the variable contains a singularity (i.e., a point at which the expression is not defined), that point is not plotted. The syntax for the **ezplot** function is described as follows:

ezplot(S) Generates a plot of **S** in the range $[-2\pi, 2\pi]$. **S** is assumed to be a function of one variable.

ezplot(S, [xmin, xmax]) Generates a plot of **S** in the range **[xmin, xmax]**. **S** is assumed to be a function of one variable.

PRACTICE!

Try out the **ezplot** statement with the following examples:

```
ezplot('cos(x)');
ezplot('cos(x)', [ -12 * pi, 12 * pi]);
ezplot('cos(x)/atan(x)',[-12 * pi, 12 * pi]);
```

The following expressions illustrate the use of these rules in determining the independent variable in symbolic expressions:

EXPRESSIONS	symvar(s)
'tan(y,x)'	x
'x^3 - 2*x^ + 3'	x
'1/(cos(x) + 2)'	x
'3*a*b - 6'	b

5.1.2   Simplification of Mathematical Expressions

A number of functions are available for simplifying mathematical expressions by **collecting coefficients**, **expanding terms**, **factoring expressions**, or just making the expression simpler. A summary of these functions is as follows:

collect(S) Collects coefficients of **S**.

collect(S,'v') Collects coefficients of **S** with respect to the independent variable **'v'**.

expand(S) Performs an expansion of **S**.

factor(S) Returns the factorization of **S**.

simple(S) Simplifies the form of **S** to a shorter form, if possible.

simplify(S) Simplifies **S** using Maple's simplification rules.

To illustrate these functions, we will use the symbolic expressions **S1**, . . . , **S5** from the previous practice session.

EXAMPLE 5.4:

Use MATLAB to perform the given actions on the symbolic expressions from the previous practice session. The results of the simplification functions are listed to the right of each example.

1.   **factor(S1)**   **(x-3)*(x+3)**
2.   **expand(S2)**   **x^2-6*x+9**

```
3.   simplify(S3)  x-5
4.   factor(S4)    (x+5)*(x-3)*(x+1)
5.   collect(S5)   6*x-11*y+13*b
```

The **simple** function attempts to simplify a symbolic expression by using several different algebraic methods. Each method is displayed, along with its results. Finally, the shortest result is chosen as the answer.

EXAMPLE 5.5:

As an example, we will use the **simple** function to find the simplest form of the following expression:

$$\frac{x^2 - 3x - 10}{x + 2}$$

First, create a symbolic expression **S3**:

```
S3 = sym('(x^2-3*x-10)/(x+2)');
```

Then call the **simple** function:

```
simple(S3)
```

The methods used and the various results are as follows:

```
simplify:          x-5
radsimp:           x-5
combine(trig):     x-5
factor:            x-5
expand:            1/(x+2)*x^2-3/(x+2)*x-10/(x+2)
combine:           (x^2-3*x-10)/(x+2)
convert(exp):      (x^2-3*x-10)/(x+2)
convert(sincos):   (x^2-3*x-10)/(x+2)
convert(tan):      (x^2-3*x-10)/(x+2)
collect(x):        (x^2-3*x-10)/(x+2)
ans =              x-5
```

5.1.3   Operations on Symbolic Expressions

The standard arithmetic operations can be applied to symbolic expressions. Addition, subtraction, multiplication, division, and raising an expression to a power are performed by using the standard arithmetic operators. This feature can best be illustrated with examples. Before proceeding with the examples, create symbolic objects **S6**, **S7**, and **S8** for the following expressions:

$$S6 = \frac{1}{y - 3}$$

$$\mathbf{S7} = \frac{3y}{y+2}$$

$$\mathbf{S8} = (y+4)(y-3)y$$

EXAMPLE 5.6:

We will multiply the symbolic objects for $\dfrac{1}{y-3}$ and $(y-4)(y-3)y$. Also, we will use the **pretty** function to display the results in typeset form.

```
pretty(S6* S8)
```

The results are as follows:

```
(y + 4)y
```

Similarly, we will raise the symbolic expression **S7** to the third power and use **pretty** to print the results:

```
pretty(S7^3)
```

The results are as follows:

```
            3
           y
     27 --------
             3
        (y + 2)
```

The **poly2sym** function converts a numerical vector to a symbolic expression. The symbolic representation is the polynomial whose coefficients are listed in the vector. The default independent variable is x, but another variable may be named as a second argument to the function. The **sym2poly** function creates a coefficient vector from the symbolic representation of a polynomial.

EXAMPLE 5.7:

First, we create a vector containing the coefficients of the polynomial:

```
V = [1 -4 0 2 45];
```

Then, we create a symbolic expression for the polynomial represented by **V**:

```
poly2sym(V)
```

The result is as follows:

```
ans =
    x^4-4*x^3+2*x+45
```

The **horner** function transposes a symbolic polynomial **S** into its nested (Horner) representation. The Horner representation of the symbolic object **S4** from the previous example,

$$x^3 + 3x^2 - 13x - 15$$

is

$$-15 + (-13 + (x + 3) * x) * x$$

EXAMPLE 5.8:   For example, to convert the polynomial

$$x^4 - 2x^2 + 5x - 15$$

to its Horner representation, we type

```
S = sym('x^4-2*x^2+5*x-15');
horner(S)
```

The result is as follows:

```
ans =
     ((x^2-2)*x+5)*x-15
```

PRACTICE!

Use MATLAB to perform the given symbolic operations. It is assumed that you have already created symbolic objects **S1**, ..., **S8** using the **sym** function.

    a.   **S6 * S8**
    b.   **S2^(1/2)**
    c.   **simple (S2^(1/2))**
    d.   **factor(S4/'(x-3)')**

5.2   EQUATION SOLVING

Symbolic math functions can be used to solve a single equation, a system of equations, and differential equations. Brief descriptions of the functions for solving a single equation or a system of equations are as follows:

solve(f)   Solves the symbolic equation **f** for its symbolic variable. Solves the equation **f=0** for its symbolic variable if **f** is a symbolic expression.

solve(f1,...fn)   Solves the system of equations represented by **f1,...,fn**.

To illustrate the use of the **solve** function, assume that the following equations have been defined:

```
eq1 = sym('x-3=4');
eq2 = sym('x^2-x-6=0');
eq3 = sym('x^2+2*x+4=0');
eq4 = sym('3*x+2*y-z=10');
eq5 = sym('-x+3*y+2*z=5');
eq6 = sym('x-y-z=-1');
```

The following table shows the resulting values from the **solve** function:

REFERENCE	FUNCTION VALUE
solve(eq1)	7
solve(eq2)	[-2, -3]'
solve(eq3)	[-1+i*3^(1/2), -1-i*3^(1/2)]'
solve(eq4,eq5,eq6)	x = -2, y = 5, z = -6

The function for solving ordinary differential equations is **dsolve**, but it is not discussed in this text. For more information on determining the symbolic solution to ordinary differential equations, see the second edition of *Engineering Problem Solving with Matlab*, by Delores M. Etter.

PRACTICE!

Solve the given systems of equations using symbolic mathematics. Compare your answers with those computed using the matrix methods from Chapter 4.

1.
$$-2x_1 + x_2 = -3$$
$$x_1 + x_2 = 3$$

2.
$$10x_1 - 7x_2 + 0x_3 = 7$$
$$-3x_1 + 2x_2 + 6x_3 = 4$$
$$5x_1 + x_2 + 5x_3 = 6$$

3.
$$x_1 + 4x_2 - x_3 + x_4 = 2$$
$$2x_1 + 7x_2 + x_3 - 2x_4 = 16$$
$$x_1 + 4x_2 - x_3 + 2x_4 = 1$$
$$3x_1 - 10x_2 - 2x_3 + 5x_4 = -15$$

5.3 DIFFERENTIATION AND INTEGRATION

The operations of **differentiation** and **integration** are used extensively in solving engineering problems. In this section, we discuss the differentiation and integration of symbolic expressions. In the next chapter, we discuss techniques for performing numerical differentiation and numerical integration using data values instead of symbolic expressions.

5.3.1 Differentiation

The **diff** function is used to determine the symbolic derivative of a symbolic expression. There are four forms in which the **diff** function can be used to perform symbolic differentiation:

diff(f)   Returns the derivative of the expression **f** with respect to the default independent variable.

diff(f,'t')   Returns the derivative of the expression **f** with respect to the variable **t**.

diff(f,n)   Returns the **n**th derivative of the expression **f** with respect to the default independent variable.

diff(f,'t',n)   Returns the **n**th derivative of the expression **f** with respect to the variable **t**.

We now present several examples using the **diff** function for symbolic differentiation. First, we define the following symbolic expressions:

```
S1 = sym('6*x^3-4*x^2+b*x-5');
S2 = sym('sin(a)');
S3 = sym('(1-t^3)/(1+t^4)');
```

The following table shows function references and their corresponding values:

REFERENCE	FUNCTION VALUE
`diff(S1)`	`18*x^2-8*x+b`
`diff(S1,2)`	`36*x-8`
`diff(S1,'b')`	`x`
`diff(S2)`	`cos(a)`
`diff(S3)`	`-3*t^2/(1+t^4)-4*(1-t^3)/(1+t^4)^2*t^3`
`simplify(diff(S3))`	`t^2*(-3+t^4-4*t)/(1+t^4)^2`

PRACTICE!

Determine the first and second derivatives of the following functions, using MATLAB's symbolic functions:

1. $g(x) = x^3 - 5x^2 + 2x + 8$
2. $g_2(x) = (x^2 + 4x + 4) * (x - 1)$
3. $g_3(x) = (x^2 - 2x + 2) / (10x - 24)$
4. $g_4(x) = (x^5 - 4x^4 - 9x^3 + 32)^2$

5.3.2   Integration

The **int** function is used to integrate a symbolic expression **f**. This function attempts to find the symbolic expression **F** such that **diff(F) = f**. It is possible that the integral (or antiderivative) may not exist in closed form or that MATLAB cannot find the integral. In such cases, the function will return the unevaluated command. The **int** function can be used in the following forms:

int(f)   Returns the integral of the expression **f** with respect to the default independent variable.

int(f,'t')   Returns the integral of the expression **f** with respect to the variable **t**.

int(f,a,b) Returns the integral of the expression **f** with respect to the default independent variable evaluated over the interval **[a,b]**, where **a** and **b** are numeric expressions.

int(f,'t',a,b) Returns the integral of the expression **f** with respect to the variable **t** evaluated over the interval **[a,b]**, where **a** and **b** are numeric expressions.

int(f,'m','n') Returns the integral of the expression **f** with respect to the default independent variable evaluated over the interval **[m,n]**, where **m** and **n** are symbolic expressions.

We now present several examples that use the **int** function for symbolic integration. First, we define the following symbolic expressions:

```
S1 = sym('6*x^3-4*x^2+b*x-5');
S2 = sym('sin(a)');
S3 = sym('sqrt(x)');
```

The following table shows function references and their corresponding values:

REFERENCE	FUNCTION VALUE
`int(S1)`	`3/2*x^4-4/3*x^3+1/2*b*x^2-5*x`
`int(S2)`	`-cos(a)`
`int(S3)`	`2/3*x^(3/2)`
`int(S3,'a','b')`	`2/3*b^(3/2)-2/3*a^(3/2)`
`int(S3,0.5,0.6)`	`2/25*15^(1/2)-1/6*2^(1/2)`
`double(int(S3, 0.5, 0.6))`	`0.0741`

PRACTICE!

Use MATLAB's symbolic functions to determine the values of the following integrals:

a. $\int_{0.5}^{0.6} |x|\, dx$

b. $\int_{0}^{1} |x|\, dx$

c. $\int_{-1}^{-0.5} |x|\, dx$

d. $\int_{-0.5}^{0.5} |x|\, dx$

SUMMARY

In this chapter, we presented MATLAB's functions for performing symbolic mathematics. Examples were given to illustrate simplification of expressions, evaluation of operations with symbolic expressions, and derivation of symbolic solutions to equations. In addition, we presented the MATLAB functions for determining the symbolic derivatives and integrals of expressions.

MATLAB SUMMARY

This MATLAB summary lists and briefly describes all of the special characters, commands, and functions that were defined in this chapter.

Special Character	
$'$	used to enclose a symbolic expression

Commands and Functions	
`collect`	collects coefficients of a symbolic expression
`diff`	differentiates a symbolic expression
`expand`	expands a symbolic expression
`ezplot`	generates a plot of a symbolic expression
`factor`	factors a symbolic expression
`findsym`	find symbolic variables in a symbolic expression
`horner`	converts a symbolic expression into a nested form
`int`	integrates a symbolic expression
`numden`	returns the numerator and denominator expressions
`poly2sym`	converts a vector to a symbolic polynomial
`pretty`	prints a symbolic expression in typeset form
`simple`	shortens a symbolic expression
`simplify`	simplifies a symbolic expression
`solve`	solves an equation
`subs`	replace variables in a symbolic expression
`sym2poly`	converts a symbolic expression to a coefficient vector

KEY TERMS

collecting coefficients
differentiation
expanding terms

factoring expressions
integration
symbolic algebra

symbolic expression

Problems

1. Create symbolic objects **S1** and **S2** for the following expressions:

$$S1 = (x - 1)^2 + 2x - 1$$
$$S2 = x$$

2. Execute the **simple** function using **S1** as an argument. Which of the nine simplification methods succeeds in finding the simplest expression for **S1**?

3. What is the result of the symbolic division **S2/S1**? What is the result of **factor(S2/S1)**?

4. Solve the equation $\dfrac{x-1}{x^2+4} = 2$.

5. Differentiate the following function: $f(x) = \dfrac{-4}{t^9}$

6. Evaluate the following integral: $\displaystyle\int_1^2 \frac{1}{y^4}\, dy$

Weather Balloons   Assume that the following polynomial represents the altitude in meters during the first 48 hours following the launch of a weather balloon:

$$h(t) = -0.12t^4 + 12t^3 - 380t^2 + 4100t + 220$$

Assume that the units of t are hours.

7.   Use MATLAB to determine the equation for the velocity of the weather balloon, using the fact that the velocity is the derivative of the altitude.

8.   Use MATLAB to determine the equation for the acceleration of the weather balloon. Compare the results obtained from using the fact that the acceleration is the second derivative of the altitude and from using the fact that the acceleration is the derivative of the velocity. (See Problem 1.)

9.   Use your answers to Problems 1 and 2 to generate plots of the altitude, velocity, and acceleration for the interval 0 to 48 hours.

Water Flow   Assume that water is pumped into an initially empty tank. It is known that the rate of flow of water into the tank at time t (in seconds) is $50 - t$ liters per second. The amount of water Q that flows into the tank during the first x seconds can be shown to be equal to the integral of the expression $(50 - t)$ evaluated from 0 to x seconds.

10.   Determine a symbolic equation that represents the amount of water in the tank after x seconds.

11.   Determine the amount of water in the tank after 30 seconds.

12.   Determine the amount of water that flowed into the tank between 10 seconds and 15 seconds after the flow was initiated.

Elastic Spring   Consider a spring with the left end held fixed and the right end free to move along the x-axis. We assume that the right end of the spring is at the origin $x = 0$ when the spring is at rest. When the spring is stretched, the right end of the spring is at some new value of x that is greater than zero. When the spring is compressed, the right end of the spring is at some value of x that is less than zero. Assume that a spring has a natural length of 1 ft and that a force of 10 lbs is required to compress the spring to a length of 0.5 ft. It can then be shown that the work (in ft/lb) done to stretch the spring from its natural length to a total length of n ft is equal to the integral of $20x$ over the interval from 0 to $n - 1$.

13.   Use MATLAB to determine a symbolic expression that represents the amount of work necessary to stretch the spring to a total length of n ft.

14.   What is the amount of work done to stretch the spring to a total of 2 ft?

15.   If the amount of work exerted is 25 ft/lb, what is the length of the stretched spring?

6

Numerical Techniques

GRAND CHALLENGE: ENHANCED OIL AND GAS RECOVERY

The design and construction of the Alaska pipeline presented numerous engineering challenges. One of the most important problems that had to be addressed was how to protect the permafrost (the perennially frozen subsoil in arctic or subarctic regions) from the heat of the pipeline itself. The oil flowing in the pipeline is warmed by pumping stations and by friction from the walls of the pipe such that the supports holding the pipeline have to be insulated or even cooled to keep them from melting the permafrost at their bases.

SECTIONS

6.1 Interpolation
6.2 Curve Fitting: Linear and Polynomial Regression
6.3 Numerical Integration
6.4 Problem Solving Applied: Pipeline Flow Analysis
6.5 Numerical Differentiation

OBJECTIVES

After reading this chapter, you should be able to

- perform linear and cubic-spline interpolation
- calculate the best-fit straight line and polynomial to a set of data points
- compute a numerical estimate for an integral
- apply the principles in this chapter to a problem in pipeline flow analysis, and
- compute a numerical estimate for a derivative.

6.1   INTERPOLATION

In this section, we present two types of interpolation: linear interpolation and cubic-spline interpolation. In both techniques, we assume that we have a set of data points that represents a set of xy-coordinates for which y is a function of x—that is, $y = f(x)$. We further assume that we need to estimate a value $f(b)$, which is not one of the original data points, but for which b is between two of the x values from the original set of data points. We want to approximate (or interpolate) a value for $f(b)$, using the information from the original set of data points. In Figure 6.1, we show a set of six data points that have been connected with straight-line segments and that have also been connected with cubic-degree polynomial segments. From this figure, we see that the values determined for the function between sample points depend on the type of interpolation that we select.

6.1.1   Linear Interpolation

Linear interpolation is one of the most common techniques for estimating data values between two given data points. If we assume that the function between the two points can be estimated by a straight line drawn between the points, we can compute the value of the function at any point between the two data values, using an equation derived from similar triangles.

6.1.2   Cubic-Spline Interpolation

A cubic spline is a smooth curve constructed to go through a set of points. The curve between each pair of points is a third-degree polynomial (which has the general form $a_0 x^3 + a_1 x^2 + a_2 x + a_3$) that is computed so that it provides a smooth curve between the two points and a smooth transition from the third-degree polynomial between the previous pair of points. Refer to the cubic spline shown in Figure 6.1, which connects six points. A total of five different cubic equations are used to generate this smooth function that joins all six points.

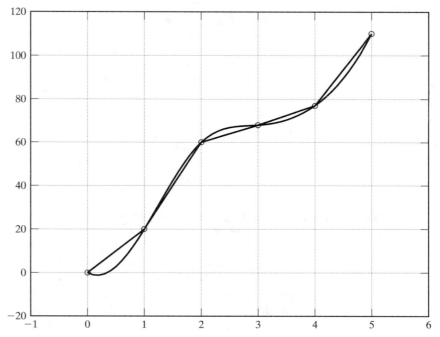

Figure 6.1.   Linear and cubic-spline interpolation.

6.1.3 **interp1** Function

The MATLAB function that performs interpolation has three forms. Each form assumes that vectors **x** and **y** contain the original data values and that another vector **x_new** contains the new points for which we want to compute interpolated **y_new** values. (The **x** values should be in ascending order, and the **x_new** values should be within the range of the **x** values.) A summary of these forms is as follows:

> **interp1(x,y,x_new)** Returns a vector the size of **x_new** that contains the interpolated **y** values that correspond to **x_new** using linear interpolation.
>
> **interp1(x,y,x_new,'linear')** Returns a vector the size of **x_new** that contains the interpolated **y** values that correspond to **x_new** using linear interpolation.
>
> **interp1(x,y,x_new,'spline')** Returns a vector the size of **x_new** that contains the interpolated **y** values that correspond to **x_new** using cubic-spline interpolation.

PRACTICE!

Use the following sets of data points for this example:

```
x = 0:5;
y = [0,20,60,68,77,110];
```

Define a finer grain of **x** values for the interpolation:

```
newx = 0:0.1:5;
```

Use the **interp1** function twice, once with linear interpolation and once with cubic–spline interpolation. Generate plots of each interpolation, as depicted in Figure 6.1. The MATLAB commands are as follows:

```
newy_1 = interp1(x,y,newx,'linear');
newy_2 = interp1(x,y,newx,'spline');
plot(newx,newy_1,newx,newy_2,x,y,'o'),
title('Linear and Cubic Spline Interpolation'),
    grid,axis([-1,6,-20,120]),pause
```

MATLAB provides other one-dimensional interpolation methods, including nearest neighbor interpolation and cubic interpolation. In addition, MATLAB provides two-dimensional **(interp2)** and three-dimensional **(interp3)** interpolation functions. These functions will not be discussed here. Type **help interp2** or **help interp3** to see a description and examples of each.

6.2 CURVE FITTING: LINEAR AND POLYNOMIAL REGRESSION

Assume that we have a set of data points collected from an experiment. After plotting the data points, we find that they generally fall in a straight line. However, if we were to try to draw a straight line through the points, only a couple of the points would probably fall exactly on the line. A **least-squares** curve-fitting method could be used to find the straight line that is the closest to the points by minimizing the distance from each point to the straight line. Although this line can be considered a "best fit" to the data points, it is possible that none of the points would actually fall on the line of best fit. (Note that this method is very different from interpolation, because the curves used in linear interpolation and cubic-spline interpolation actually contained all of the original data points.) In this section, we first discuss fitting a straight line to a set of data points, and then we discuss fitting a polynomial to a set of data points.

6.2.1 Linear Regression

Linear regression is the name given to the process that determines the linear equation which is the best fit to a set of data points, in terms of minimizing the sum of the squared distances between the line and the data points. To understand this process, we first consider the set of data values used in the discussion on interpolation from the previous section. If we plot these points, it appears that a good estimate of a line through the points is $y = 20x$, as shown in Figure 6.2. The following commands were used to generate this plot:

```
%   These statements compare a linear model
%   with a set of data points. (Figure 6.2)
%
x = 0:5;
y = [0,20,60,68,77,110];
y1 = 20*x;
plot(x,y1,x,y,'o'),title('Linear Estimate'),
xlabel('Time, s'),ylabel('Temperature, Degrees F'),grid,
axis([-1,6,-20,120)],pause
```

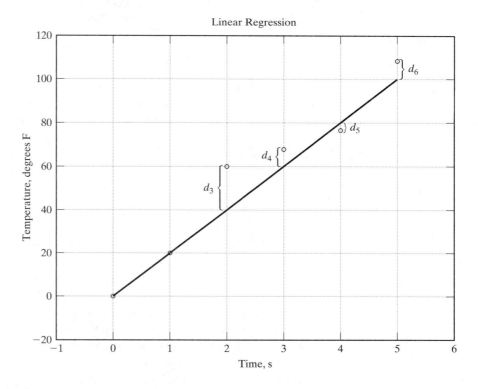

Figure 6.2. A linear estimate.

To measure the quality of the fit of this linear estimate to the data, we first determine the distance from each point to the linear estimate; these distances are also shown in Figure 6.2. The first two points fall exactly on the line, so d_1 and d_2 are zero. The value of d_3 is equal to $60 - 40$, or 20; the rest of the distances can be computed in a similar way. If we compute the sum of the distances, some of the positive and negative values would cancel each other and give a sum that is smaller than it should be. To avoid this problem, we could add absolute values or squared values; linear regression uses squared values. Therefore, the measure of the quality of the fit of this linear estimate is the sum of the **squared distances** between the points and the linear estimates. This sum can be computed with the following command:

```
sum_sq = sum((y-y1).^2)
```

For this set of data, the value of **sum_sq** is 573.

If we drew another line through the points, we could compute the sum of squares that corresponds to this new line. Of the two lines, the better fit is provided by the line with the smaller sum of squared distances. To find the line with the smallest sum of squared distances, we can write an equation that computes the distances using a general linear equation, $y = mx + b$. We then write an equation that represents the sum of the squared distances; this equation will have m and b as its variables. Using techniques from calculus, we can compute the derivatives of the equation with respect to m and b and set the derivatives equal to zero. The values of m and b that are determined in this way represent the straight line with the minimum sum of squared distances. The MAT-LAB statement for computing this best-fit linear equation is discussed in the next section. For the data presented in this section, the best fit is shown in Figure 6.3; the corresponding sum of squares is 356.8190.

6.2.2   Polynomial Regression

In the previous discussion, we presented a technique for computing the linear equation that best fits a set of data. A similar technique can be developed using a single polynomial (not a set of polynomials, as in a cubic spline) to fit the data by minimizing the distance of the polynomial from the data points. First, recall that a polynomial with one variable can be written by using the following general formula:

$$f(x) = a_0 x^n + a_1 x^{n-1} + a_2 x^{n-2} + \cdots + a_{n-1}x + a_n$$

The **degree of a polynomial** is equal to the largest value used as an exponent. Therefore, the general form of a cubic polynomial is

$$g(x) = a_0 x^3 + a_1 x^2 + a_2 x + a_3$$

Note that a linear equation is also a polynomial of degree one.

In Figure 6.4 we plot the original set of data points that we used in the linear regression example, along with plots of the best-fit polynomials with degrees two through five. Note that as the degree of the polynomial increases, the number of points that fall on the curve also increases. If a set of n points is used to determine an nth-degree polynomial, all n points will fall on the polynomial.

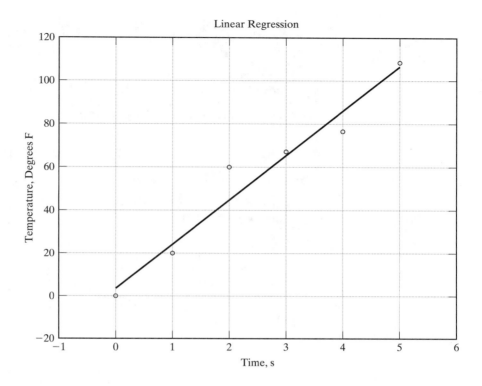

Figure 6.3. Data and best-fit line.

6.2.3 `polyfit` and `polyval` Functions

The MATLAB function for computing the best fit to a set of data with a polynomial with a specified degree is the **polyfit** function. This function has three arguments: the x and y coordinates of the data points and the degree n of the polynomial. The function returns the coefficients, in descending powers of x, of the nth-degree polynomial that fits the vectors **x** and **y**. (Note that an nth-degree polynomial has $n + 1$ coefficients.) A summary of this function is

> **polyfit(x,y,n)**   Returns a vector of **n+1** coefficients that represents the best-fit polynomial of degree **n** for the **x** and **y** coordinates. The coefficient order corresponds to decreasing powers of **x**.

The best linear fit for a set of data can be found using the **polyfit** function with $n = 1$. The following example demonstrates how to find a best-fit linear estimate.

PRACTICE!

First, we define six data points:

```
x = 0:5;
y = [0,20,60,68,77,110];
```

The **polyfit** function (with $n = 1$) returns a vector containing the coefficients of a polynomial of degree one:

```
coef = polyfit(x,y,1)
coef = 20.8286    3.7619
```

The best y estimates are calculated using the coefficients:

```
ybest = coeff(1) * x + coeff(2)
ybest = 3.7619  24.5905  45.4190  66.2476  87.0762  107.9048
```

We can calculate the sum of squares using the following commands:

```
sum_sq = sum((y - ybest).^2)
sum_sq = 356.8190
```

The best-fit linear equation is plotted as follows:

```
plot(x,ybest,x,y,'o'),title('Linear Regression'),
xlabel('Time, s'),ylabel('Temperature, Degrees F'),grid,
axis([-1,6,-20,120]),pause
```

This plot is depicted in Figure 6.3.

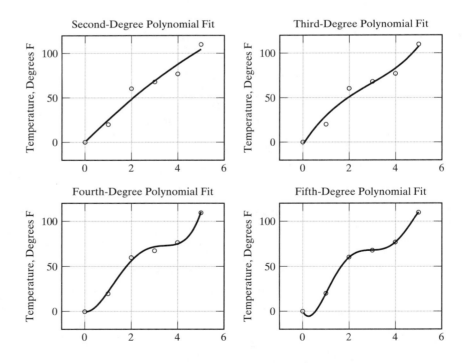

Figure 6.4. Polynomial fits.

The **polyval** function is used to evaluate a polynomial at a set of data points. The first argument of the **polyval** function is a vector containing the coefficients of the polynomial (in an order corresponding to decreasing powers of x), and the second argument is the vector of x values for which we want polynomial values. A summary of the function is

polyval(coef,x)   Returns a vector of polynomial values $f(x)$ that correspond to the **x** vector values. The order of the coefficients corresponds to decreasing powers of x.

In the previous example, we computed the points of the linear regression using values from the coefficients. We could also have computed them using the **polyval** function:

```
ybest = polyval(coef,x);
```

The **polyfit** and **polyval** functions can be used in combination to write concise programs that find best-fit polynomials. This method is convenient if you are using higher order polynomials. For example, the following statement returns an evaluation of the best-fit fifth-order polynomial using the data in the previous example:

```
polyval(polyfit(x,y,5),0:0.05:5)
```

To plot the results in a single statement, use the following command:

```
plot(0:0.05:5, polyval(polyfit(x,y,5),0:0.05:5))
```

PRACTICE!

This example illustrates the computation of the best-fit polynomials of degree one through degree five. We will use the same data that were used for the linear regression example.

First, define the six data points:

```
x = 0:5;
y = [0, 20, 60, 68, 77, 110];
```

Define a finer grain of **x** values for the polynomial evaluation:

```
newx = 0:0.05:5;
```

The following loop generates a 5 × 101 matrix that contains the evaluation of the best-fit polynomials from order one to five.

```
for n=1:5
   f(:, n) = polyval( polyfit(x, y, n), newx)';
end
```

Then, for example, to plot the third-degree polynomial, we could use the following statement:

```
plot(newx, f(:,3), x, y, 'o')
```

Additional statements are necessary to define the subplot, label each plot, and set the limits of the axes. The results are depicted in Figure 6.4.

From the previous discussion on polynomial fits, we would expect that the lower degree polynomials would not contain all of the data points and that the fifth-degree polynomial would contain all six data points. The plots in Figure 6.4 verify these expectations.

PRACTICE!

The following set of data represents the flow of water through a culvert as a function of the water's height.

HEIGHT, FT	FLOW, CFS
1.70	2.60
1.95	3.60
2.60	4.03
2.92	6.45
4.04	11.22
5.24	30.61

1. Compute a best-fit linear equation for the data. Evaluate the resulting linear equation for $x = 0:0.05:6$. Plot the results.
2. Compute a third-order best-fit polynomial for the data. Evaluate the resulting polynomial for $x = 0:0.05:6$. Plot the results.

6.3 NUMERICAL INTEGRATION

The integral of a function $f(x)$ over the interval $[a,b]$ is defined to be the area under the curve of $f(x)$ between a and b, as shown in Figure 6.5. If the value of this integral is K, the notation to represent the integral of $f(x)$ between a and b is

$$K = \int_a^b f(x)dx$$

For many functions, this integral can be computed analytically. However, for a number of functions, the integral cannot easily be computed analytically and thus requires a numerical technique to estimate its value. The numerical evaluation of an integral is also called quadrature, a term that comes from an ancient geometrical problem.

The numerical integration techniques estimate the function $f(x)$ by another function $g(x)$, where $g(x)$ is chosen so that we can easily compute the area under $g(x)$. Then, the better the estimate of $g(x)$ to $f(x)$, the better will be the estimate of the integral of $f(x)$. Two of the most common numerical integration techniques estimate $f(x)$ with a set of piecewise linear functions or with a set of piecewise parabolic functions. If we estimate the function with piecewise linear functions, we can then compute the area of the trapezoids that compose the area under the piecewise linear functions; this technique is called the **trapezoidal rule**. If we estimate the function with piecewise quadratic functions, we can then compute and add the areas of these components; this technique is called **Simpson's rule**.

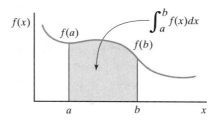

Figure 6.5. Integral of $f(x)$ from a to b.

6.3.1 Trapezoidal Rule and Simpson's Rule

If the area under a curve is represented by trapezoids and if the interval $[a,b]$ is divided into n equal sections, then the area can be approximated by the formula (trapezoidal rule)

$$K_T = \frac{b-a}{2n}(f(x_0) + 2f(x_1) + 2f(x_2) + \ldots + 2f(x_{n-1}) + f(x_n))$$

where the x_i values represent the end points of the trapezoids and where $x_0 = a$ and $x_n = b$.

If the area under a curve is represented by areas under quadratic sections of a curve and if the interval $[a,b]$ is divided into $2n$ equal sections, then the area can be approximated by the formula (Simpson's rule)

$$K_S = \frac{h}{3}(f(x_0) + 4f(x_1) + 2f(x_2) + 4f(x_3) + \ldots + 2f(x_{2n-2}) + 4f(x_{2n-1}) + f(x_{2n}))$$

where the x_i values represent the end points of the sections and where $x_0 = a$, $x_{2n} = b$, and $h = (b - a)/(2n)$.

If the piecewise components of the approximating function are higher degree functions (the trapezoidal rule uses linear functions, and Simpson's rule uses quadratic functions), the integration techniques are referred to as **Newton-Cotes integration techniques**.

The estimate of an integral improves as we use more components (such as trapezoids) to approximate the area under a curve. If we attempt to integrate a function with a singularity (a point at which the function or its derivatives are infinity or are not defined), we may not be able to get a satisfactory answer with a numerical integration technique.

6.3.2 MATLAB Quadrature Functions

MATLAB has two quadrature functions for performing numerical function integration. The **quad** function uses an adaptive form of Simpson's rule, whereas **quadl** uses an adaptive Lobatto quadrature. The **quadl** function is better at handling functions with certain types of singularities, such as $\int_0^1 \sqrt{x}\, dx$. Both functions print a warning message if they detect a singularity, but an estimate of the integral is still returned.

The simplest form of the **quad** and **quadl** functions requires three arguments. The first argument is the name (in quotes) of the MATLAB function that returns a vector of values of $f(x)$ when given a vector of input values **x**. This function name can be the name of another MATLAB function, such as **sin**, or it can be the name of a user-written

MATLAB function. The second and third arguments are the integral limits **a** and **b**. A summary of these functions is as follows:

quad('function',a,b)   Returns the area of the **'function'** between **a** and **b**, assuming that **'function'** is a MATLAB function.

quadl('function',a,b)   Returns the area of the **'function'** between **a** and **b**, assuming that **'function'** is a MATLAB function.

PRACTICE!

The script shown next can be used to compare the results of the **quad** and **quad8** functions with the analytically calculated results. The script prompts the user for a specified interval.

```
%  These statements compare the quad and quad8 functions
%  with the analytical results for the integration of the
%  square root of x over an interval [a,b], where a and b
%  are nonnegative.
%
a = input(' Enter left endpoint (nonnegative): ');
b = input('Enter right endpoint (nonnegative): ');
%
%  k is the computed analytical result
k = (2/3)*(b^(1.5) - a^(1.5));
%
%  The following two statements compute the quad and quad
%  functions from a to b
kquad=quad('sqrt',a,b);
kquadl=quadl('sqrt',a,b);
%
%  Display the results
fprintf('Analytical: %f \n',k);
fprintf('      Quad: %f \n',kquad);
fprintf('      Quadl: %f \n',kquadl);
```

These integration techniques can handle some singularities that occur at one or the other interval end points, but they cannot handle singularities that occur within the interval. For these cases, you should consider dividing the interval into subintervals and providing estimates of the singularities using other results, such as l'Hôpital's Rule.

To illustrate, assume that we want to determine the integral of the square-root function for nonnegative values of a and b:

$$K_Q = \int_a^b \sqrt{x}\, dx$$

The square-root function $f(x) = \sqrt{x}$ is plotted in Figure 6.6 for the interval [0, 5]; the values of the function are complex for $x < 0$. This function can be integrated analytically to yield the following for nonnegative values of a and b:

$$K = \frac{2}{3}(b^{3/2} - a^{3/2})$$

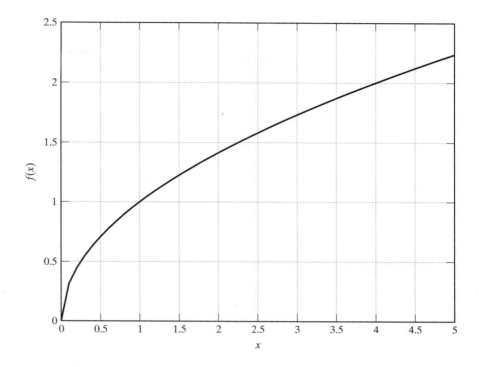

Figure 6.6. Square-root function.

PRACTICE!

You can cut and paste the script for this example into a file and test it. If you select an interval that contains a singularity you will see a message similar to the following:

```
Recursion level limit reached in quad.   Singularity likely.
```

The following examples demonstrate the script's use:

```
Enter left endpoint (nonnegative): 1.5
Enter right endpoint (nonnegative): 15
Analytical: 37.505089
    Quad: 37.504990
    Quadl: 37.505088
Enter left endpoint (nonnegative): 0.2
Enter right endpoint (nonnegative): 5
Analytical: 7.393931
    Quad: 7.393905
    Quadl: 7.393926
```

The **quad** and **quadl** functions can also include a fourth argument, which represents a tolerance. If the tolerance is omitted, a default value of 0.001 is assumed. The integration function continues to refine its estimate for the integration until the relative error is less than the tolerance, using the following iterative test:

$$\frac{\text{previous estimate} - \text{current estimate}}{\text{previous estimate}} < \text{tolerance}$$

PRACTICE!

Sketch the function $f(x) = x$, and indicate the areas specified by the given integrals. Then compute the integrals by hand, and compare your results with those generated by the **quad** function.

1. $\int_{0.5}^{0.6} |x| \, dx$ 3. $\int_{0}^{1} |x| \, dx$

2. $\int_{-1}^{-0.5} |x| \, dx$ 4. $\int_{-0.5}^{0.5} |x| \, dx$

6.4   PROBLEM SOLVING APPLIED: PIPELINE FLOW ANALYSIS

In this section, we perform computations in an application related to the enhanced oil recovery grand challenge. The friction in a circular pipeline causes a velocity profile to develop in the flowing oil. Oil that is in contact with the walls of the pipe does not move at all, whereas oil at the center of the flow moves the fastest. The diagram in Figure 6.7 shows how the velocity of the oil varies across the diameter of the pipe and defines the variables used in this analysis. The following equation describes this velocity profile:

$$v(r) = v_{max} \left(\overline{1} - \frac{r}{r^0} \right)^{\frac{1}{n}}$$

The variable v is an integer between 5 and 10 that defines the shape of the forward flow of the oil. In this case, the value of v for the diagram in Figure 6.7 is 8. The average flow velocity of the pipe can be computed by integrating the velocity profile from zero to the pipe's radius, r_0. Thus, we have

$$v_{ave} = \frac{\int_{0}^{r_0} v(r) 2\pi dr}{\pi r_0^2} = \frac{2v_{max}}{r_0^2} \int_{0}^{r_0} r\overline{1} - \frac{r}{r_0} \sqrt{1/2} dr$$

The values of v_{max} and v can be measured experimentally, and the value of r_0 is the radius of the pipe. We will assume that v_{max} is 1.5 m, is 0.5 m, and v is 8.

First, we will plot the function $\overline{1} - \frac{r}{r_0} \sqrt{1/2} \, dr$ for r, varying from 0 to 0.5 meters, in increments of 0.01 meters. We can approximate the area under this function using a triangle and a trapezoid, as shown in Figure 6.8.

$$\text{area} = 0.1 \leftrightarrow 0.35 + \frac{0.4 \leftrightarrow 0.35}{2} = 0.105 \, \text{m}^2$$

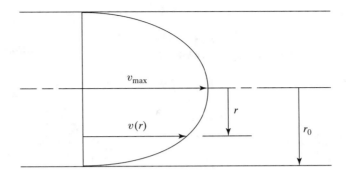

Figure 6.7. Velocity profile of flowing oil.

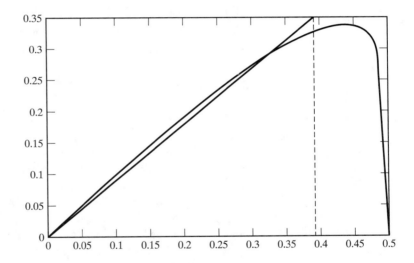

Figure 6.8. Function approximation using a triangle and rectangle.

Now we will compute the area, using the **quad** and **quad8** functions. In order to use the **quad** and **quads** functions, we first create a function that describes the velocity profile:

```
function s=vel(x)
%
% vel - describes the velocity profile
%
s = x.* (1-(x/0.5)).^(1/8);
```

Then we use this function as an argument to the **quad** and **quad8** functions:

```
% plot velocity profile, compare quad and quad8 solutions
r = [0.0:0.01:0.5];
plot(r, vel(r));
fprintf('Quad = %f\n',quad('vel',0.0,0.5));
fprintf('Quadl = %f\n',quadl('vel',0.0,0.5));
```

Compare the answers with the foregoing estimation:

```
Quad = 0.104563
Quadl = 0.104574
```

The average flow velocity for this pipe can now be calculated (using the **quadl** estimate):

$$v_{max} = \frac{2v_{max}}{r_0^2} \int_0^{r_0} r\bar{1} - \frac{r}{r_0} \sqrt{1/(2dr)} = \left(\frac{2 \leftrightarrow 1.5}{(0.5)} \leftrightarrow 0.1045 \right)$$

6.5   NUMERICAL DIFFERENTIATION

The derivative of a function $f(x)$ is defined to be a function $f'(x)$ that is equal to the rate of change of $f(x)$ with respect to x. The derivative can be expressed as a ratio, with the change in $f(x)$ indicated by $df(x)$ and the change in x indicated by dx, giving

$$f'(x) = \frac{df(x)}{dx}$$

There are many physical processes for which we want to measure the rate of change of a variable. For example, velocity is the rate of change of position (as in meters per second), and acceleration is the rate of change of velocity (as in meters per second squared). It can also be shown that the integral of acceleration is velocity and that the integral of velocity is position. Hence, integration and differentiation have a special relationship, in that they can be considered to be inverses of each other: The derivative of an integral returns the original function, and the integral of a derivative returns the original function, to within a constant value.

The derivative $f'(x)$ can be described graphically as the slope of the function $f(x)$, where the slope of $f(x)$ is defined to be the slope of the tangent line to the function at the specified point. Thus, the value of $f'(x)$ at the point a is $f'(a)$, and it is equal to the slope of the tangent line at the point a, as shown in Figure 6.9.

Because the derivative of a function at a point is the slope of the tangent line at the point, a value of zero for the derivative of a function at the point x_k indicates that the line is horizontal at that point. Points with derivatives of zero are called **critical points** and can represent either a horizontal region of the function or a local maximum or a local minimum of the function. (The point may also be the global maximum or global minimum, as shown in Figure 6.10, but more analysis of the entire function would be needed to determine this.) If we evaluate the derivative of a function at several points in an interval and we observe that the sign of the derivative changes, then a local maximum or a local minimum occurs in the interval. The second derivative (the derivative of $f'(x)$) can be used to determine whether or not the critical points represent local maxima or local minima. More specifically, if the second derivative of an **extrema point** is positive, then the value of the function at the extrema point is a local minimum; if the second derivative of an extrema point is negative, then the value of the function at the extrema point is a local maximum.

6.5.1 Difference Expressions

Numerical differentiation techniques estimate the derivative of a function at a point x_k by approximating the slope of the tangent line at x_k using values of the function at points near x_k. The approximation of the slope of the tangent line can be done in several ways, as shown in Figure 6.11.

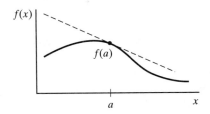

Figure 6.9. Derivative of $f(x)$ at $x = a$.

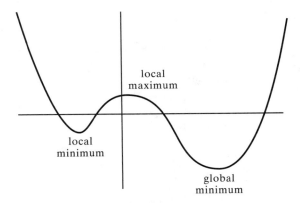

Figure 6.10. Example of function with critical points.

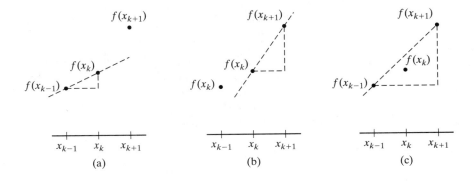

Figure 6.11. Techniques for computing $f'(x_k)$.

Figure 6.11(a) assumes that the derivative at x_k is estimated by computing the slope of the line between $f(x_{k-1})$ and $f(x_k)$, as in

$$f'(x_k) = \frac{f(x_k) - f(x_{k-1})}{x_k - x_{k-1}}$$

This type of derivative approximation is called a **backward difference approximation**.

Figure 6.11(b) assumes that the derivative at x_k is estimated by computing the slope of the line between $f(x_k)$ and $f(x_{k+1})$, as in

$$f'(x_k) = \frac{f(x_{k+1}) - f(x_k)}{x_{k+1} - x_k}$$

This type of derivative approximation is called a **forward difference approximation**.

Figure 6.11(c) assumes that the derivative at x_k is estimated by computing the slope of the line between $f(x_{k-1})$ and $f(x_{k+1})$, as in

$$f'(x_k) = \frac{f(x_{k+1}) - f(x_{k-1})}{x_{k+1} - x_{k-1}}$$

This type of derivative approximation is called a **central difference approximation**, and we usually assume that x_k is halfway between x_{k-1} and x_{k+1}. The quality of all of these types of derivative computations depends on the distance between the points used to estimate the derivative; the estimate of the derivative improves as the distance between the two points decreases.

The second derivative of a function $f(x)$ is the derivative of the first derivative of the function:

$$f''(x) = \frac{df'(x)}{dx}$$

This function can be evaluated using slopes of the first derivative. Thus, if we use backward differences, we have

$$f''(x_k) = \frac{f'(x_k) - f'(x_{k-1})}{x_k - x_{k-1}}$$

Similar expressions can be derived for computing estimates of higher derivatives.

6.5.2 **diff** Function

The **diff** function computes differences between adjacent values in a vector, generating a new vector with one fewer value. If the **diff** function is applied to a matrix, it operates on the columns of the matrix as if each column were a vector. A second, optional argument specifies the number of times to recursively apply **diff**. Each time **diff** is applied, the length of the vector is reduced in size. A third, optional argument specifies the dimensions in which to apply the function. The forms of **diff** are summarized as follows:

> **diff(X)** For a vector **X**, **diff** returns
> **[X(2)-X(1) X(3)-X(2) ... X(n)-X(n-1)]**.

diff(X)   For a matrix **X**, **diff** returns the matrix of column differences
[X(2:m,:) - X(1:m-1,:)]

diff(X,n,dim)   The general form of **diff** returns the **n**th difference
function along dimension **dim** (a scalar). If $n >=$ the length of **dim**, then
diff returns an empty array.

PRACTICE!

To illustrate, define vectors **x**, **y**, and **z** as follows:

```
x = [0 1 2 3 4 5];
y = [2 3 1 5 8 10];
z = [1 3 5; 1 5 10];
```

Then the vector generated by **diff(x)** is

```
diff(x)
ans =
     1      1      1      1      1
```

The vector generated by **diff(y)** is

```
diff(y)
ans =
     1     -2      4      3      2
```

The next example recursively executes **diff** twice. Note that the length of the
returned vector is 4:

```
diff(y,2)
ans =
    -3      6     -1     -1
```

The **diff** function can be applied to either dimension of matrix **z**:

```
diff(z,1,1)
ans =
     0      2      5
diff(z,1,2)
ans =
     2      2
     4      5
```

An approximate derivative dy can be computed by using **diff(y)./diff(x)**.

Note that these values of dy are correct for both the forward difference equation
and the backward difference equation. The distinction between the two methods for
computing the derivative is determined by the values of the vector **xd**, which correspond to the derivative dy. If the corresponding values of **xd** are [1,2,3,4,5], dy computes a backward difference. If the corresponding values of **xd** are [0,1,2,3,4], dy
computes a forward difference.

PRACTICE!

As an example, consider the function given by the following polynomial:

$$f(x) = x^5 - 3x^4 - 11x^3 + 27x^2 + 10x - 24$$

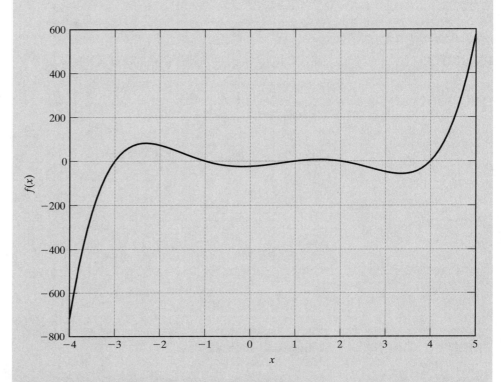

Figure 6.12. Fifth-degree polynomial.

A plot of this function is shown in Figure 6.13. Recall that the zeros of the derivative correspond to the points of local minima or local maxima of a function. The function in this example does not have a global minimum or global maximum, because the function ranges from -∞ to ∞. The local minima and maxima (or critical points) of this function occur at -2.3, -0.2, 1.5, and 3.4. You can use the **find** function to identify the critical points of a function. Assume that we want to compute the derivative of this function over the interval [4,5]. We can perform this operation using the **diff** function as, shown in the following script, where *df* represents **df** and *xd* represents the *x* values corresponding to the derivative:

```
%  Evaluate f(x) and f'(x).
%
x = -4:0.1:5;
f = x.^5 - 3*x.^4 - 11*x.^3 + 27*x.^2 + 10*x - 24;
df = diff(f)./diff(x);
xd = x(2:length(x));
plot(xd, df);
```

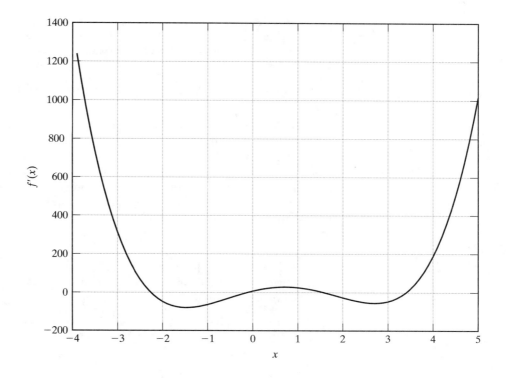

Figure 6.13. Derivative of fifth-degree polynomial.

PRACTICE!

In this example, we will use the values of **df** and **xd** from the previous example. The **find** function is used to determine the indices k of the locations in **product** for which **df(k)** is equal to 0. These indices are then used with the vector **xd** to print the approximation to the locations of the critical points.

```
%  Find locations of critical points of f'(x).
%
product = df(1:length(df)-1).*df(2:length(df));
critical = xd(find(product<0))

critical =
-2.3000   -0.2000    1.5000    3.4000
```

PRACTICE!

This example shows the method for computing a central difference derivative using vectors **x** and **f**.

```
x = -4:0.1:5;
f = x.^5 - 3*x.^4 - 11*x.^3 + 27*x.^2 + 10*x - 24;
```

The following script evaluates $f(x)$ using central differences:

```
%  Evaluate f'(x) using central differences.
%
```

```
numerator = f(3:length(f)) - f(1:length(f)-2);
denominator = x(3:length(x)) - x(1:length(x)-2);
dy = numerator./denominator;
xd = x(2:length(x)-1);
```

You may want to plot **dy** and compare it with Figure 6.13.

In the example discussed in this section, we assumed that we had the equation of the function to be differentiated, and thus we could generate points of the function. In many engineering problems, the data to be differentiated are collected from experiments. Thus, we cannot choose the points to be close together to get a more accurate measure of the derivative. In these cases, it might be a good solution to use the techniques from Section 6.2. These techniques allow us to determine an equation for a polynomial that fits a set of data and then compute points from the equation to use in computing values of the derivative.

PRACTICE!

For each of the given functions, plot the function, its first derivative, and its second derivative over the interval $[-10,10]$. Then use MATLAB commands to print the locations of the local minima.

1. $g_1(x) = x^3 - 5x^2 + 2x + 8$

2. $g_2(x) = x^2 + 4x + 4$

3. $g_3(x) = x^2 - 2x + 2$

4. $g_4(x) = 10x - 24$

5. $g_5(x) = x^5 - 4x^4 - 9x^3 + 32x^2 + 28x - 48$

SUMMARY

In this chapter, we explained the difference between interpolation and least-squares curve fitting. Two types of interpolation were presented: linear interpolation and cubic-spline interpolation. After presenting the MATLAB commands for performing these types of interpolations, we then turned to least-squares curve fitting using polynomials. This discussion explained how to determine the best fit to a set of data using a polynomial with a specified degree and then how to use the best-fit polynomial to generate new values of the function. Techniques for numerical integration and differentiation were also presented in this chapter. Numerical integration techniques approximate the area under a curve, and numerical differentiation techniques approximate the slope of a curve. The functions for integration are **quad** and **quadl**. The function used to compute the derivative of a function is the **diff** function, which computes differences between adjacent elements of a vector.

MATLAB SUMMARY

This MATLAB summary lists and briefly describes all of the commands and functions that were defined in this chapter.

Commands and Functions

diff	computes the differences between adjacent values
interp1	computes linear and cubic interpolation

Commands and Functions

`polyfit`	computes a least-squares polynomial
`polyval`	evaluates a polynomial
`quad`	computes the integral under a curve (Simpson)
`quadl`	computes the integral under a curve (Lobatto)

KEY TERMS

approximation	degree of a polynomial	linear interpolation
backwards difference	derivative	linear regression
central difference approximation	extrema points	quadrature
critical points	forward difference approximation	Simpson's rule
cubic spline	least-squares	trapezoidal rule

Problems

1. Generate $f(x) = x^2$ for $x = [-3\ -1\ 0\ 2\ 5\ 6]$.

   - Compute and plot the linear and cubic-spline interpolation of the data points over the range $[-3:0.05:6]$.
   - Compute the value of $f(4)$ using linear interpolation and cubic-spline interpolation. What are the respective errors when the answer is compared with the actual value of $f(4)$?

2. **Cylinder Head Temperatures.** Assume that the following set of temperature measurements is taken from the cylinder head in a new engine that is being tested for possible use in a race car:

TIME, s	TEMPERATURE, °F
0.0	0.0
1.0	20.0
2.0	60.0
3.0	68.0
4.0	77.0
5.0	110.0

   a. Compare plots of these data, assuming linear interpolation and assuming cubic interpolation for values between the data points, using time values from 0 to 5 in increments of 0.1 s.

   b. Using the data from part (a), find the time value for which there is the largest difference between its linear-interpolated temperature and its cubic-interpolated temperature.

c. Assume that we measure temperatures at three points around the cylinder head in the engine, instead of at just one point. The set of data is then the following:

TIME, s	TEMP1	TEMP2	TEMP3
0.0	0.0	0.0	0.0
1.0	20.0	25.0	52.0
2.0	60.0	62.0	90.0
3.0	68.0	67.0	91.0
4.0	77.0	82.0	93.0
5.0	110.0	103.0	96.0

Assume that these data have been stored in a matrix with six rows and four columns. Determine interpolated values of temperature at the three points in the engine at 2.6 seconds, using linear interpolation.

d. Using the information from part (c), determine the time that the temperature reached 75 degrees at each of the three points in the cylinder head.

3. **Spacecraft Accelerometer.** The guidance and control system for a spacecraft often uses a sensor called an *accelerometer*, which is an electromechanical device that produces an output voltage proportional to the applied acceleration. Assume that an experiment has yielded the following set of data:

ACCELERATION	VOLTAGE
4	0.593
2	0.436
0	0.061
2	0.425
4	0.980
6	1.213
8	1.646
10	2.158

a. Determine the linear equation that best fits this set of data. Plot the data points and the linear equation.

b. Determine the sum of the squares of the distances of these points from the line of best fit determined in part (a).

c. Compare the error sum from part (b) with the same error sum computed from the best quadratic fit. What do these sums tell you about the two models for the data?

4. Compute $\tan(x)$ for $x = [-1:0.05:1]$. Compute the best-fit polynomial of order four that approximates $\tan(x)$. Plot $\tan(x)$ and the generated polynomial on the same graph. What is the sum of square error of the polynomial approximation for the data points in x?

5.   **Function Analysis.** Let the function f be defined by the following equation:

$$f(x) = 4e^{-x}$$

Plot this function over the interval [0,1]. Use numerical integration techniques to estimate the integral of (x) over [0,0.5] and over [0,1].

6.   **Sounding Rocket Trajectory.** The following data set represents the time and altitude values for a sounding rocket that is performing high-altitude atmospheric research on the ionosphere:

TIME, s	ALTITUDE, m
0	60
10	2,926
20	10,170
30	21,486
40	33,835
50	45,251
60	55,634
70	65,038
80	73,461
90	80,905
100	87,368
110	92,852
120	97,355
130	100,878
140	103,422
150	104,986
160	106,193
170	110,246
180	119,626
190	136,106
200	162,095
210	199,506
220	238,775
230	277,065
240	314,375
250	350,704

a.   Plot the altitude data. The velocity function is the derivative of the altitude function. Using numerical differentiation, compute the velocity values from these data, using a backward difference. Plot the velocity data. (Note that the rocket is a two-stage rocket.)

b. The acceleration function is the derivative of the velocity function. Using the velocity data determined from part (a), compute the acceleration data, using a backward difference. Plot the acceleration data.

7. **Simple Root Finding.** Even though MATLAB makes it easy to find the roots of a function, sometimes all that is needed is a quick estimate. This can be done by plotting a function and zooming in very close to see where the function equals zero. Since MATLAB draws straight lines between data points in a plot, it is good to draw circles or stars at each data point, in addition to the straight lines connecting the points. Plot the following function, and zoom in to find the roots:

```
n = 5;
x = linspace(0,2*pi,n);
y = x .* sin(x) + cos(1/2*x).^2 - 1./(x - 7);
plot (x,y,'-o')
```

Increase the value of **n** to increase the accuracy of the estimate.

Consider the data points in the following two vectors:

$$\mathbf{X} = [0.1 \ 0.3 \ 5.0 \ 6.0 \ 23.0 \ 24.0]$$
$$\mathbf{Y} = [2.8 \ 2.6 \ 18.1 \ 26.8 \ 486.1 \ 530.0]$$

8. Determine the best-fit polynomial of order 2 for the data. Calculate the sum of squares for your results. Plot the best-fit polynomial for the six data points.

9. Generate a new **X** containing 250 uniform data points in increments of 0.1 from [0.1, 25.0]. Using the best-fit polynomial coefficients from the previous problem, generate a new **Y** containing 250 data points. Plot the results.

10. Compute an estimate of the derivative using the new **X** and new **Y** generated in the previous problem. Compute the coefficients of the derivative. Plot the derivative.

Index